PROBLEMS AND SOLUTIONS FOR
COMPLEX ANALYSIS

T0202419

Springer
New York
Berlin
Heidelberg
Barcelona
Hong Kong
London
Milan
Paris
Singapore
Tokyo

Rami Shakarchi

PROBLEMS AND SOLUTIONS FOR

COMPLEX ANALYSIS

With 46 Illustrations

Springer

Rami Shakarchi
Department of Mathematics
Princeton University
Princeton, NJ 08544-1000
USA

Mathematics Subject Classification (2000): 30-01

Shakarchi, Rami.
    Problems and solutions for Complex analysis / Rami Shakarchi.
       p.   cm.
    "Contains all the exercises and solutions of Serge Lang's Complex
analysis""—Pref.
    Includes bibliographical references.
    ISBN 0-387-98831-9 (softcover : alk. paper)
    1. Mathematical analysis—Problems, exercises, etc.   2. Functions
of complex variables—Problems, exercises, etc.   I. Lang, Serge,
1927–   . Complex analysis.   II. Title.
QA301.S48   1999
515´.9—dc21   99-13255

Printed on acid-free paper.

Production managed by Lesley Poliner; manufacturing supervised by Jerome Basma.
Photocomposed copy prepared from the author's TeX files.
Printed and bound by Sheridan Books, Ann Arbor, MI.
Printed in the United States of America.

9 8 7 6 5 4 3

ISBN 0-387-98831-9 Springer-Verlag New York Berlin Heidelberg

*This book is dedicated to my parents,
Mohamed and Mireille Shakarchi,
in appreciation for their love and support.*

Preface

This book contains all the exercises and solutions of Serge Lang's *Complex Analysis*. Chapters I through VIII of Lang's book contain the material of an introductory course at the undergraduate level and the reader will find exercises in all of the following topics: power series, Cauchy's theorem, Laurent series, singularities and meromorphic functions, the calculus of residues, conformal mappings and harmonic functions. Chapters IX through XVI, which are suitable for a more advanced course at the graduate level, offer exercises in the following subjects: Schwarz reflection, analytic continuation, Jensen's formula, the Phragmen–Lindelöf theorem, entire functions, Weierstrass products and meromorphic functions, the Gamma function and the Zeta function. This solutions manual offers a large number of worked out exercises of varying difficulty.

I thank Serge Lang for teaching me complex analysis with so much enthusiasm and passion, and for giving me the opportunity to work on this answer book. Without his patience and help, this project would be far from complete.

I thank my brother Karim for always being an infinite source of inspiration and wisdom. Finally, I want to thank Mark McKee for his help on some problems and Jennifer Baltzell for the many years of support, friendship and complicity.

Rami Shakarchi
Princeton, New Jersey
1999

Contents

I

Complex Numbers and Functions

I.1   Definition

Exercise I.1.1. *Express the following complex numbers in the form* $x + iy$, *where* x, y *are real numbers.*
(a) $(-1 + 3i)^{-1}$
(b) $(1 + i)(1 - i)$
(c) $(1 + i)i(2 - i)$
(d) $(i - 1)(2 - i)$
(e) $(7 + \pi i)(\pi + i)$
(f) $(2i + 1)\pi i$
(g) $(\sqrt{2}i)(\pi + 3i)$
(h) $(i + 1)(i - 2)(i + 3)$

Solution. (a) $\frac{-1}{10} - \frac{3}{10}i$. (b) 2. (c) $-1 + 3i$. (d) $-1 + 3i$. (e) $6\pi + i(7 + \pi^2)$. (f) $-2\pi + i\pi$. (g) $-3\sqrt{2} + \pi\sqrt{2}i$. (h) $-8 - 6i$.

Exercise I.1.2. *Express the following complex numbers in the form* $x + iy$, *where* x, y *are real numbers.*
(a) $(1 + i)^{-1}$
(b) $\frac{1}{3+i}$
(c) $\frac{2+i}{2-i}$
(d) $\frac{1}{2-i}$
(e) $\frac{1+i}{i}$
(f) $\frac{i}{1+i}$
(g) $\frac{2i}{3-i}$
(h) $\frac{1}{-1+i}$

Solution. (a) $\frac{1}{2} - \frac{1}{2}i$. (b) $\frac{3}{10} - \frac{1}{10}i$. (c) $\frac{3}{5} + \frac{4}{5}i$. (d) $\frac{2}{5} + \frac{1}{5}i$. (e) $1 - i$. (f) $\frac{1}{2} + \frac{1}{2}i$. (g) $\frac{-1}{5} + \frac{3}{5}i$. (h) $\frac{-1}{2} - \frac{1}{2}i$.

Exercise I.1.3. *Let* α *be a complex number* $\neq 0$. *What is the absolute value of* $\alpha/\bar{\alpha}$? *What is* $\bar{\bar{\alpha}}$?

Solution. Let $\alpha = a + ib$. Then

$$\frac{\alpha}{\bar{\alpha}} = \frac{a + ib}{a - ib} = \frac{a^2 - b^2 + 2abi}{a^2 + b^2}$$

so

$$\left|\frac{\alpha}{\overline{\alpha}}\right|^2 = \frac{(a^2 - b^2)^2 + 4a^2b^2}{(a^2 + b^2)^2} = \frac{(a^2 + b^2)^2}{(a^2 + b^2)^2} = 1.$$

Moreover,

$$\overline{\overline{\alpha}} = \overline{a - ib} = a + ib = \alpha.$$

Exercise I.1.4. *Let α, β be two complex numbers. Show that $\overline{\alpha\beta} = \overline{\alpha}\overline{\beta}$ and that*

$$\overline{\alpha + \beta} = \overline{\alpha} + \overline{\beta}.$$

Solution. Suppose $\alpha = a + ib$ and $\beta = c + id$. Then

$$\overline{\alpha}\overline{\beta} = (a - ib)(c - id) = ac - bd - i(ad + bc) = \overline{\alpha\beta},$$

and

$$\overline{\alpha} + \overline{\beta} = a - ib + c - id = (a + c) - i(b + d) = \overline{\alpha + \beta}.$$

Exercise I.1.5. *Justify the assertion made in the proof of Theorem 1.2, that the real part of a complex number is $\leq$ its absolute value.*

Solution. Suppose $z = x + iy$. Then

$$x^2 \leq x^2 + y^2,$$

and taking square roots we obtain the inequality

$$|\operatorname{Re}(z)| \leq |z|.$$

Exercise I.1.6. *If $\alpha = a + ib$ with a, b real, then b is called the **imaginary part** of α and we write $b = \operatorname{Im}(\alpha)$. Show that $\alpha - \overline{\alpha} = 2i \operatorname{Im}(\alpha)$. Show that*

$$\operatorname{Im}(\alpha) \leq |\operatorname{Im}(\alpha)| \leq |\alpha|.$$

Solution. We have

$$\alpha - \overline{\alpha} = a + ib - a + ib = 2ib.$$

The first inequality is obvious, while the second inequality follows from

$$\operatorname{Im}(\alpha)^2 \leq \operatorname{Re}(\alpha)^2 + \operatorname{Im}(\alpha)^2.$$

Exercise I.1.7. *Find the real and imaginary parts of $(1 + i)^{100}$.*

Solution. Since $(1+i)^2 = 2i$ we have $(1+i)^{100} = 2^{50}i^{50}$. But $i^{50} = (-1)^{25} = -1$ so

$$(1 + i)^{100} = -2^{50}.$$

Exercise I.1.8. *Prove that for any two complex numbers z, w we have:*
(a) $|z| \leq |z - w| + |w|$
(b) $|z| - |w| \leq |z - w|$
(c) $|z| - |w| \leq |z + w|$

Solution. All three inequalities are obtained by writing $z = z - w + w$ and applying the triangle inequality.

Exercise I.1.9. *Let $\alpha = a + ib$ and $z = x + iy$. Let c be real > 0. Transform the condition*

$$|z - \alpha| = c$$

into an equation involving only x, y, a, b, and c, and describe in a simple way what geometric figure is represented by this equation.

Solution. By definition we see that $|z - \alpha| = c$ is equivalent to

$$\sqrt{(x - a)^2 + (y - b)^2} = c,$$

so the above equation describes the circle of radius c centered at α.

Exercise I.1.10. *Describe geometrically the sets of points z satisfying the following conditions.*
(a) $|z - i + 3| = 5$
(b) $|z - i + 3| > 5$
(c) $|z - i + 3| \leq 5$
(d) $|z + 2i| \leq 1$
(e) $\operatorname{Im} z > 0$
(f) $\operatorname{Im} z \geq 0$
(g) $\operatorname{Re} z > 0$
(h) $\operatorname{Re} z \geq 0$

Solution. (a) Circle of radius 5 centered at $i - 3$.
(b) Complement of the closed disc of radius 5 centered at $i - 3$.
(c) Closed disc of radius 5 centered at $i - 3$.
(d) Closed disc of radius 1 centered at $-2i$.
(e) Open upper half plane.
(f) Closed upper half plane.
(g) Open right half plane.
(h) Closed right half plane.

I.2   Polar Form

Exercise I.2.1. *Put the following complex numbers in polar form.*
(a) $1 + i$
(b) $1 + i\sqrt{2}$
(c) -3
(d) $4i$
(e) $1 - i\sqrt{2}$
(f) $-5i$
(g) -7
(h) $-1 - i$

Solution. (a) $\sqrt{2}e^{\frac{i\pi}{4}}$ (b) Let $\theta \in [0, 2\pi)$ be the angle such that $\cos\theta = 1/\sqrt{3}$ and $\sin\theta = \sqrt{2}/\sqrt{3}$. Then $1 + i\sqrt{2} = \sqrt{3}e^{i\theta}$. (c) $3e^{i\pi}$ (d) $4e^{\frac{i\pi}{2}}$ (e) If θ is as in (b), then $1 - i\sqrt{2} = \sqrt{3}e^{i(2\pi-\theta)}$ (f) $5e^{\frac{3i\pi}{2}}$ (g) $7e^{i\pi}$ (h) $\sqrt{2}e^{\frac{5i\pi}{4}}$.

Exercise I.2.2. *Put the following complex numbers in the ordinary form $x + iy$.*
(a) $e^{3i\pi}$
(b) $e^{2i\pi/3}$
(c) $3e^{i\pi/4}$
(d) $\pi e^{-i\pi/3}$
(e) $e^{2i\pi/6}$
(f) $e^{-i\pi/2}$
(g) $e^{-i\pi}$
(h) $e^{-5i\pi/4}$

Solution. (a) -1 (b) $-\frac{1}{2} + i\frac{\sqrt{3}}{2}$ (c) $\frac{3}{\sqrt{2}} + i\frac{3}{\sqrt{2}}$ (d) $\frac{\pi}{2} - i\frac{\pi\sqrt{3}}{2}$ (e) $\frac{1}{2} + i\frac{\sqrt{3}}{2}$ (f) $-i$ (g) -1 (h) $-\frac{1}{\sqrt{2}} + i\frac{1}{\sqrt{2}}$

Exercise I.2.3. *Let α be a complex number $\neq 0$. Show that there are two distinct complex numbers whose square is α.*

Solution. Suppose $\alpha = re^{i\varphi}$. Then the two solutions to $z^2 = \alpha$ are $\sqrt{r}e^{i(\varphi/2)}$ and $\sqrt{r}e^{i(\pi+\varphi/2)}$. See Exercise 6.

Exercise I.2.4. *Let $a + bi$ be a complex number. Find real numbers x, y such that*

$$(x + iy)^2 = a + ib,$$

expressing x, y in terms of a and b.

Solution. Since $(x + iy)^2 = x^2 - y^2 + 2ixy$ we have $x^2 - y^2 = a$ and $2xy = b$. Taking absolute values we also get $x^2 + y^2 = \sqrt{a^2 + b^2}$. These three equations imply

$$x^2 = \frac{a + \sqrt{a^2 + b^2}}{2} \quad \text{and} \quad y^2 = \frac{\sqrt{a^2 + b^2} - a}{2}.$$

and therefore

$$x = \sqrt{\frac{a + \sqrt{a^2 + b^2}}{2}} \quad \text{and} \quad y = (\text{sign } b)\sqrt{\frac{\sqrt{a^2 + b^2} - a}{2}}$$

solves our problem.

Exercise I.2.5. *Plot all the complex numbers z such that $z^n = 1$ on a sheet of graph paper, for $n = 2, 3, 4,$ and 5.*

Solution. The equation $z^2 = 1$ has two solutions, 1 and -1 which we plot as stars. The equations $z^3 = 1$ has three solutions, 1, $e^{2\pi i/3}$ and $e^{4\pi i/3}$ which we plot as dots.

The equation $z^4 = 1$ has four solutions, 1, i, -1 and $-i$ which we plot as stars. The equation $z^5 = 1$ has five solutions, 1, $e^{2\pi i/5}$, $e^{4\pi i/5}$, $e^{6\pi i/5}$ and $e^{8\pi i/5}$ which we plot as dots. (See figure at top of next page.)

Exercise I.2.6. *Let α be a complex number $\neq 0$. Let n be a positive integer. Show that there are n distinct complex numbers z such that $z^n = \alpha$. Write these complex numbers in polar form.*

Solution. We use the expressions $z = re^{i\theta}$ and $\alpha = se^{i\varphi}$. Then the equation $z^n = \alpha$ is equivalent to

$$z^n = r^n e^{ni\theta} = se^{i\varphi},$$

hence

$$\frac{r^n}{s}e^{i(n\theta-\varphi)} = 1.$$

Taking absolute values we get $r^n = s$. Moreover $n\theta - \varphi$ must be an integral multiple of 2π, so the set of solutions of the equation $z^n = \alpha$ is

$$S = \left\{ s^{1/n}e^{i\left(\frac{\varphi}{n}\right)}, s^{1/n}e^{i\left(\frac{\varphi}{n}+\frac{2\pi}{n}\right)}, \ldots, s^{1/n}e^{i\left(\frac{\varphi}{n}+(n-1)\frac{2\pi}{n}\right)} \right\}.$$

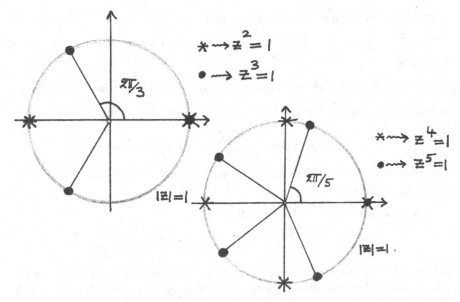

Exercise I.2.7. *Find the real and imaginary parts of $i^{1/4}$, taking the fourth root such that its angle lies between 0 and $\pi/2$.*

Solution. Since $i = e^{i\pi/2}$ we see that we want to find the real and imaginary parts of $e^{i\pi/8}$. Hence

$$\operatorname{Re}(i^{1/4}) = \cos\frac{\pi}{8} = \sqrt{\frac{1+\cos\frac{\pi}{4}}{2}} = \sqrt{\frac{1}{2} + \frac{\sqrt{2}}{4}}$$

and

$$\operatorname{Im}(i^{1/4}) = \sin\frac{\pi}{8} = \sqrt{\frac{1-\cos\frac{\pi}{4}}{2}} = \sqrt{\frac{1}{2} - \frac{\sqrt{2}}{4}}.$$

Exercise I.2.8. *(a) Describe all complex numbers z such that $e^z = 1$.*
(b) Let w be a complex number. Let α be a complex number such that $e^\alpha = w$. Describe all complex numbers z such that $e^z = w$.

Solution. (a) If $z = x + iy$ where x and y are real, then $e^z = 1$ is equivalent to $e^x e^{iy} = 1$. Taking absolute values we see that $e^x = 1$ hence $x = 0$ and therefore y must be an integral multiple of 2π. Conversely, any complex number of the form $2\pi ki$ with $k \in \mathbf{Z}$ is a solution of $e^z = 1$.
(b) The number w is nonzero because $e^\alpha = w$ and therefore

$$e^{z-\alpha} = \frac{e^z}{e^\alpha} = \frac{w}{w} = 1$$

hence by (a) we must have $\operatorname{Re}(z) = \operatorname{Re}(\alpha)$ and $\operatorname{Im}(z) - \operatorname{Im}(\alpha)$ must be an integral multiple of 2π. We can also express the solutions of $e^z = w$ independently of α.

Writing $w = re^{i\theta}$ and $z = x + iy$ we see that

$$x = \log r \quad \text{and} \quad y \equiv \theta \pmod{2\pi}.$$

Conclude.

Exercise I.2.9. *If* $e^z = e^w$, *show that there is an integer* k *such that* $z = w + 2\pi k i$.

Solution. Writing z and w in the form $a + ib$ we find after taking absolute values that

$$e^{\operatorname{Re}(z)} = e^{\operatorname{Re}(w)} \quad \text{and} \quad e^{i\operatorname{Im}(z)} = e^{i\operatorname{Im}(w)}$$

therefore $\operatorname{Re}(z) = \operatorname{Re}(w)$ and $\operatorname{Im}(z) - \operatorname{Im}(w)$ is equal to an integral multiple of 2π, as was to be shown.

Exercise I.2.10. *(a) If* θ *is real, show that*

$$\cos\theta = \frac{e^{i\theta} + e^{-i\theta}}{2} \quad \text{and} \quad \sin\theta = \frac{e^{i\theta} - e^{-i\theta}}{2i}.$$

(b) For arbitrary complex z, *suppose we define* $\cos z$ *and* $\sin z$ *by replacing* θ *with* z *in the above formula. Show that the only values of* z *for which* $\cos z = 0$ *and* $\sin z = 0$ *are the usual real values from trigonometry.*

Solution. (a) The two formulas follow from

$$e^{i\theta} + e^{-i\theta} = \cos\theta + i\sin\theta + \cos\theta - i\sin\theta = 2\cos\theta$$

and

$$e^{i\theta} - e^{-i\theta} = \cos\theta + i\sin\theta - \cos\theta + i\sin\theta = 2i\sin\theta.$$

(b) If $\cos z = 0$ then by definition we get $e^{iz} + e^{-iz} = 0$. Multiplying this equation by e^{iz} and writing $z = x + iy$ we obtain

$$-1 = e^{2i(x+iy)} = e^{-2y}e^{2ix}.$$

Taking absolute values we get $y = 0$ and therefore $x \equiv \pi/2 \pmod{\pi}$ as was to be shown. The equation $\sin z = 0$ is equivalent to $e^{iz} - e^{-iz} = 0$ hence $e^{2iz} = 1$. Letting $z = x + iy$ and arguing as we did before, we find that $x = 0$ and $y \equiv 0 \pmod{\pi}$.

Exercise I.2.11. *Prove that for any complex number* $z \neq 1$ *we have*

$$1 + z + \cdots + z^n = \frac{z^{n+1} - 1}{z - 1}.$$

Solution. This formula follows from

$$(z - 1)(1 + z + \cdots + z^n) = z + z^2 + \cdots + z^n + z^{n+1} - 1 - z - \cdots - z^n$$
$$= z^{n+1} - 1.$$

Exercise I.2.12. *Using the preceding exercise, and taking real parts, prove:*

$$1 + \cos\theta + \cos 2\theta + \cdots + \cos n\theta = \frac{1}{2} + \frac{\sin[(n + 1/2)\theta]}{2\sin(\theta/2)}$$

for $0 < \theta < 2\pi$.

Solution. The real part of the sum $\sum_{k=0}^{n}(e^{i\theta})^k$ is $1+\cos\theta+\cos 2\theta+\cdots+\cos n\theta$. The formula of the preceding exercise gives

$$\sum_{k=0}^{n}(e^{i\theta})^k = \frac{e^{i(n+1)\theta}-1}{e^{i\theta}-1} = \frac{e^{i(n+1/2)\theta}-e^{-i\theta/2}}{e^{i\theta/2}-e^{-i\theta/2}}$$

$$= \frac{e^{i(n+1/2)\theta}-e^{-i\theta/2}}{2i\sin(\theta/2)}$$

$$= i\frac{e^{-i\theta/2}-e^{i(n+1/2)\theta}}{2\sin(\theta/2)}.$$

The imaginary part of $e^{-i\theta/2}-e^{i(n+1/2)\theta}$ is $\sin(-\theta/2)-\sin[(n+1/2)\theta]$, whence

$$1+\cos\theta+\cos 2\theta+\cdots+\cos n\theta = \frac{\sin(\theta/2)+\sin[(n+1/2)\theta]}{2\sin\theta/2},$$

and the desired formula drops out.

Exercise I.2.13. *Let z, w be two complex numbers such that $\bar{z}w \neq 1$. Prove that*

$$\left|\frac{z-w}{1-\bar{z}w}\right| < 1 \quad \textit{if } |z| < 1 \textit{ and } |w| < 1,$$

$$\left|\frac{z-w}{1-\bar{z}w}\right| = 1 \quad \textit{if } |z| = 1 \textit{ or } |w| = 1.$$

(There are many ways of doing this. One way is as follows. First check that you may assume that z is real, say $z = r$. For the first inequality you are reduced to proving

$$(r-w)(r-\bar{w}) < (1-rw)(1-r\bar{w}).$$

You can then use elementary calculus, differentiating with respect to r and seeing what happens for $r = 0$ and $r < 1$, to conclude the proof.)

Solution. If $z = re^{i\theta}$ we make the substitution $w \leftrightarrow we^{i\theta}$ and we find

$$\left|\frac{re^{i\theta}-we^{i\theta}}{1-re^{-i\theta}we^{i\theta}}\right| = \left|\frac{r-w}{1-rw}\right|,$$

so we may assume that z is a real number $0 \le r \le 1$. We want to show that

$$(r-w)(r-\bar{w}) \le (1-rw)(1-r\bar{w})$$

with $=$ if and only if $r = 1$ or $|w| = 1$. We expand both sides, make the necessary cancellations and move all the terms to the right to see that the above inequality is equivalent to

$$0 \le (1-r^2)(1-w\bar{w}),$$

hence

$$0 \le (1-r^2)(1-|w|^2).$$

Conclude.

I.3   Complex Valued Functions

Exercise I.3.1. *Let $f(z) = 1/z$. Describe what f does to the inside and outside of the unit circle, and also what it does to points on the unit circle. This map is called **inversion** through the unit circle.*

Solution. The inversion is defined on the complex plane minus the origin. If $z = re^{i\theta}$ then

$$\frac{1}{z} = \frac{1}{r}e^{-i\theta}.$$

Let $V = \mathbf{C} - \overline{D}$, in other words, V is the complement of the closed unit disc. From the above formula we find we see that the image of D under the inversion is V and conversely, the image of V under the inversion is D.

If $r = 1$, then $1/z = e^{-i\theta}$ so the image of the unit circle is the unit circle.

Exercise I.3.2. *Let $f(z) = 1/\bar{z}$. Describe f in the same manner as in Exercise 1. This map is called **reflection** through the unit circle.*

Solution. The reflection is defined on the complex plane minus the origin. If $z = re^{i\theta}$, then

$$f(z) = \frac{1}{r}e^{i\theta}$$

so if $V = \mathbf{C} - \overline{D}$, then we see that $f(D - \{0\}) = V$ and $f(V) = D - \{0\}$.

If $r = 1$, then $f(z) = e^{i\theta}$ so the image of the unit circle is the unit circle.

Exercise I.3.3. *Let $f(z) = e^{2\pi iz}$. Describe the image under f of the set shaded in the figure on the facing page, consisting of those points $x + iy$ with $-\frac{1}{2} \le x \le \frac{1}{2}$ and $y \ge B$.*

Solution. If $z = x + iy$, then

$$f(z) = e^{2\pi i(x+iy)} = e^{-2\pi i}e^{2\pi ix}.$$

But $-\frac{1}{2} \le x \le \frac{1}{2}$ so $-\pi \le 2\pi x \le \pi$ and $y \ge B$ so $-2\pi y \le -2\pi B$. From the above expression we see that the absolute value of $f(z)$ is $e^{-2\pi y}$ and the argument of $f(z)$ is $2\pi x$. So the image of the shaded region under f is the closed disc of radius $e^{-2\pi B}$ minus the origin.

Exercise I.3.4. *Let $f(z) = e^z$. Describe the image under f of the following sets:*
(a) The set of $z = x + iy$ such that $x \le 1$ and $0 \le y \le \pi$.
(b) The set of $z = x + iy$ such that $0 \le y \le \pi$ (no condition on x).

Solution. If $z = x + iy$, then

$$f(z) = e^{x+iy} = e^xe^{iy}.$$

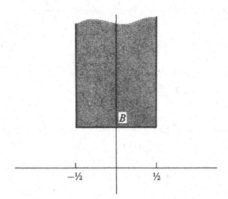

So the absolute value of $f(z)$ is e^x and the argument of $f(z)$ is y.

(a) The image of the given set is the closed upper half disc of radius e minus the origin.

(b) The image of the given region is the closed upper half plane minus the origin.

I.4   Limits and Compact Sets

Exercise I.4.1. *Let α be a complex number of absolute value < 1. What is $\lim_{n\to\infty} \alpha^n$? Proof?*

Solution. We write $\alpha = re^{i\theta}$ with $0 \le r = |\alpha| < 1$. Then

$$|\alpha^n| = |r^n e^{ni\theta}| = r^n,$$

and since $\lim_{n\to\infty} r^n = 0$ we conclude that

$$\lim_{n\to\infty} |\alpha^n| = 0,$$

and therefore

$$\lim_{n\to\infty} \alpha^n = 0.$$

Exercise I.4.2. *If $|\alpha| > 1$, does $\lim_{n\to\infty} \alpha^n$ exist? Why?*

Solution. We write $\alpha = re^{i\theta}$ with $|\alpha| = r$. Then

$$|\alpha^n| = r^n$$

so $|\alpha^n| \to \infty$ as $n \to \infty$ so the limit

$$\lim_{n\to\infty} \alpha^n$$

does not exist.

Exercise I.4.3. *Show that for any complex number $z \neq 1$, we have*

$$1 + z + \cdots + z^n = \frac{z^{n+1} - 1}{z - 1}.$$

If $|z| < 1$, show that

$$\lim_{n \to \infty} (1 + z + \cdots + z^n) = \frac{1}{1 - z}.$$

Solution. We have

$$(z - 1)(1 + z + \cdots + z^n) = z + \cdots + z^n + z^{n+1} - 1 - z - \cdots - z^n$$
$$= z^{n+1} - 1,$$

so if $z \neq 1$ the first formula drops out. If $|z| < 1$, then

$$\lim_{n \to \infty} z^{n+1} = 0,$$

so

$$\lim_{n \to \infty} (1 + z + \cdots + z^n) = \frac{1}{1 - z}$$

as was to be shown.

Exercise I.4.4. *Let f be the function defined by*

$$f(z) = \lim_{n \to \infty} \frac{1}{1 + n^2 z}.$$

Show that f is the characteristic function of the set $\{0\}$, that is, $f(0) = 1$, and $f(z) = 0$ if $z \neq 0$.

Solution. We clearly have $f(0) = 1$. If $z \neq 0$, then $|z| \neq 0$ and for all large n we have

$$|n^2 z + 1| \geq n^2 |z| - 1 \geq \frac{1}{2} n^2 |z|,$$

so for all sufficiently large n we have

$$\left| \frac{1}{n^2 z + 1} \right| \leq \frac{2}{n^2 |z|}.$$

Therefore $f(z) = 0$ as was to be shown.

Exercise I.4.5. *For $|z| \neq 1$ show that the following limit exists:*

$$f(z) = \lim_{n \to \infty} \left(\frac{z^n - 1}{z^n + 1} \right).$$

Is it possible to define $f(z)$ when $|z| = 1$ in such a way to make f continuous?

Solution. Suppose $|z| < 1$, then

$$\left|\frac{z^n - 1}{z^n + 1} - (-1)\right| = \left|\frac{2z^n}{z^n + 1}\right| \to 0$$

as $n \to \infty$, so $f(z) = -1$. If $|z| > 1$ then

$$\left|\frac{z^n - 1}{z^n + 1} - 1\right| = \frac{2}{|z^n + 1|} \to 0$$

as $n \to \infty$, so $f(z) = 1$. From these results we see that we cannot define $f(z)$ when $|z| = 1$ so as to make f continuous.

Exercise I.4.6. *Let*

$$f(z) = \lim_{n \to \infty} \frac{z^n}{1 + z^n}.$$

(a) What is the domain of definition of f, that is, for which complex numbers z does the limit exist?
(b) Give explicitly the values of $f(z)$ for the various z in the domain of f.

Solution. If $|z| < 1$, then $z^n \to 0$ as $n \to \infty$ so $f(z) = 0$. If $|z| > 1$, then $f(z) = 1$ because

$$\left|\frac{z^n}{z^n + 1} - 1\right| = \frac{1}{|z^n + 1|} \to 0$$

as $n \to \infty$.

We now investigate what happens on the unit circle. Let $z = e^{i\theta}$ with $0 \le \theta < 2\pi$. Then $1 + z^n = 1 + e^{ni\theta}$, so if $\theta = 0$ we immediately get $f(z) = 1/2$. If $\theta \ne 0$ then

$$f(z) = \frac{e^{ni\theta}}{1 + e^{ni\theta}} = \frac{1}{1 + e^{-ni\theta}},$$

and since $e^{-ni\theta}$ goes around the circle we cannot define f at the points $z = e^{i\theta}$ with $\theta \ne 0$. So if Ω denotes the unit circle minus the point 1, we see that the domain of definition of f is the set $\mathbf{C} - \Omega$.

Exercise I.4.7. *Show that the series*

$$\sum_{n=1}^{\infty} \frac{z^{n-1}}{(1 - z^n)(1 - z^{n+1})}$$

converges to $1/(1 - z)^2$ for $|z| < 1$ and to $1/z(1 - z)^2$ for $|z| > 1$. Prove that the convergence is uniform for $|z| \le c < 1$ in the first case, and $|z| \ge b > 1$ in the second. [Hint: Multiply and divide each term by $1 - z$, and do a partial fraction decomposition, getting a telescoping effect.]

Solution. (i) Let $U_n = z^n/(1 - z^n)(1 - z^{n+1})$ and let $D(z) = (1 - z^n)(1 - z^{n+1})$. Then

$$U_n = \frac{z^n(1 - z)}{D(z)(1 - z)} = \frac{1}{1 - z}\left[\frac{z^n - z^{n+1}}{D(z)}\right]$$

$$= \frac{1}{1-z} \left[\frac{(z_n - 1) + (1 - z^{n+1})}{(1 - z^n)(1 - z^{n+1})} \right]$$

$$= \frac{1}{1-z} \left[-\frac{1}{1 - z^{n+1}} + \frac{1}{1 - z^n} \right].$$

We get a telescopic sum, whence

$$\sum_{k=1}^{n} U_k = \frac{1}{1-z} \left[\frac{1}{1-z} - \frac{1}{1 - z^{n+1}} \right]$$

and therefore

$$S_n(z) = \sum_{k=1}^{n} \frac{z^{k-1}}{(1 - z^k)(1 - z^{k+1})} = \frac{1}{z(1-z)} \left[\frac{1}{1-z} - \frac{1}{1 - z^{n+1}} \right].$$

If $|z| < 1$, then $1/(1 - z^{n+1}) \to 1$ as $n \to \infty$ and therefore $S_n(z) \to 1/(1-z)^2$ as $n \to \infty$. If $|z| > 1$, then $1/(1 - z^{n+1}) \to 0$ as $n \to \infty$ so $S_n(z) \to 1/z(1 - z)^2$.
(ii) Suppose $|z| \leq c < 1$. A little algebra and part (i) imply that

$$\left| S_n(z) - \frac{1}{(1 - z)^2} \right| = \left| \frac{1}{z(1 - z)} \right| \left| \frac{z^{n+1}}{1 - z^{n+1}} \right|.$$

But $|1 - z^{n+1}| \geq 1 - |z|^{n+1} \geq 1 - c^{n+1}$, so we get the estimate

$$\left| S_n(z) - \frac{1}{(1 - z)^2} \right| \leq \frac{1}{1 - c} \frac{c^n}{1 - c^{n+1}}$$

for all z in the region $|z| \leq c < 1$. Now $c^n \to 0$ hence the convergence is uniform in the region $|z| \leq c < 1$.
   If $|z| \geq b > 1$, then

$$\left| S_n(z) - \frac{1}{z(1 - z)^2} \right| = \left| \frac{1}{z(1 - z)} \right| \left| \frac{1}{1 - z^{n+1}} \right| \leq \frac{1}{b(b - 1)} \frac{1}{b^{n+1} - 1}$$

and $b^{n+1} \to \infty$, so the convergence is uniform in the region $|z| \geq b > 1$.

I.6   The Cauchy–Riemann Equations

Exercise I.6.1. *Prove in detail that if u, v satisfy the Cauchy–Riemann equations, then the function*

$$f(z) = f(x + iy) = u(x, y) + iv(x, y)$$

is holomorphic.

Solution. The Cauchy–Riemann equations are

$$\frac{\partial u}{\partial x} = \frac{\partial v}{\partial y} \quad \text{and} \quad \frac{\partial u}{\partial y} = -\frac{\partial v}{\partial x}.$$

We use the notation of the section. Let $\zeta = \frac{\partial u}{\partial x} - i\frac{\partial u}{\partial y}$. Then using the Cauchy–Riemann equations we find that $f(z+w) - f(z) - \zeta w$ is equal to

$$\left[u(x+h, y+k) - u(x, y) - \frac{\partial u}{\partial x}h - \frac{\partial u}{\partial y}k \right]$$
$$+ i\left[v(x+h, y+k) - v(x, y) - \frac{\partial v}{\partial x}h - \frac{\partial v}{\partial y}k \right].$$

which we can rewrite as

$$|(h, k)|\sigma_1(h, k) + i|(h, k)|\sigma_2(h, k)$$

where $\lim_{(h,k)\to(0,0)} \sigma_i(h, k) = 0$ for $i = 1, 2$, because both u and v are differentiable. For $w \neq 0$ near zero we let

$$\sigma(w) = \frac{|w|}{w}(\sigma_1(w) + i\sigma_2(w)).$$

Then

$$f(z+w) - f(z) - \zeta w = w\sigma(w)$$

where $\lim_{w\to 0} \sigma(w) = 0$. This proves that f is holomorphic at z.

II

Power Series

II.1 Formal Power Series

Exercise II.1.1. *Give the terms of order ≤ 3 in the power series:*
(a) $e^z \sin z$
(b) $(\sin z)(\cos z)$
(c) $\frac{e^z - 1}{z}$
(d) $\frac{e^z - \cos z}{z}$

(e) $\frac{1}{\cos z}$
(f) $\frac{\cos z}{\sin z}$
(g) $\frac{\sin z}{\cos z}$
(h) $e^z / \sin z$

Solution. (a) $e^z \sin z = z + z^2 + \left(-\frac{1}{3!} + \frac{1}{2!}\right) z^3 + \text{higher terms.}$
(b) $(\sin z)(\cos z) = z + \left(-\frac{1}{2!} - \frac{1}{3!}\right) z^3 + \text{higher terms.}$
(c) $\frac{e^z - 1}{z} = 1 + \frac{z}{2!} + \frac{z^2}{3!} + \frac{z^3}{4!} + \text{higher terms.}$
(d) $\frac{e^z - \cos z}{z} = 1 + \left(\frac{1}{2} + \frac{1}{2}\right) z + \frac{1}{3!} z^2 + \text{higher terms.}$
(e) $\frac{1}{\cos z} = 1 + \frac{1}{2} z^2 + \left(-\frac{1}{24} + \frac{1}{4}\right) z^4 + \text{higher terms.}$
(f) $\frac{\cos z}{\sin z} = z + \left(\frac{1}{2} - \frac{1}{3!}\right) z^4 + \text{higher terms.}$
(g) $\frac{\sin z}{\cos z} = z + \left(\frac{1}{2} - \frac{1}{3!}\right) z^4 + \text{higher terms.}$
(h) $e^z / \sin z = \frac{1}{z} + 1 + \left(\frac{1}{3!} + \frac{1}{2!}\right) z + \frac{1}{3!} z^2 + \left(\frac{1}{(3!)^2} - \frac{1}{5!} + \frac{1}{3!2!} + \frac{1}{4!}\right) z^3 + \text{higher}$
terms.

Exercise II.1.2. *Let* $f(z) = \sum a_n z^n$. *Define* $f(-z) = \sum a_n(-z)^n = \sum a_n(-1)^n z^n$. *We define* $f(z)$ *to be* **even** *if* $a_n = 0$ *for n odd. We define* $f(z)$ *to be* **odd** *if* $a_n = 0$ *for n even. Verify that f is even if and only if $f(-z) = f(z)$ and f is odd if and only if $f(-z) = -f(z)$.*

Solution. Suppose f is even. Since $a_n = 0$ for n odd and $(-1)^n = 1$ if n is even, we get

$$f(-z) = \sum a_n(-1)^n z^n = \sum_{n \text{ even}} a_n(-1)^n z^n = \sum_{n \text{ even}} a_n z^n = f(z).$$

Conversely, suppose that $f(-z) = f(z)$. This implies that

$$\sum(-1)^n a_n z^n = \sum a_n z^n$$

hence $2\sum_{n \text{ odd}} a_n z^n = 0$ which implies that $a_n = 0$ for all n odd.

When f is even, a similar argument proves the desired statement.

Exercise II.1.3. *Define the **Bernoulli numbers** B_n by the power series*

$$\frac{z}{e^z - 1} = \sum_{n=0}^{\infty} \frac{B_n}{n!} z^n.$$

Prove the recursion formula

$$\frac{B_0}{n!0!} + \frac{B_1}{(n-1)!1!} + \cdots + \frac{B_{n-1}}{1!(n-1)!} = \begin{cases} 1 & \text{if } n = 1, \\ 0 & \text{if } n > 1. \end{cases}$$

Then $B_0 = 1$. Compute B_1, B_2, B_3, B_4. Show that $B_n = 0$ if n is odd $\neq 1$.

Solution. We know that $e^z = \sum_{n=0}^{\infty} z^n/n!$, so

$$\frac{z}{e^z - 1} = \frac{z}{\sum_{n=1}^{\infty} z^n/n!} = \frac{1}{\sum_{n=0}^{\infty} z^n/(n+1)!}.$$

Therefore by definition of the Bernoulli numbers we have

$$1 = \left(\sum_{n=0}^{\infty} z^n/(n+1)!\right)\left(\sum_{n=0}^{\infty} \frac{B_n}{n!} z^n\right).$$

Since we are multiplying power series we can use the formula given in the text so that

$$\sum_{k=0}^{n} \frac{B_k}{k!(n-k+1)!} = \begin{cases} 1 & \text{if } n = 0, \\ 0 & \text{if } n > 0. \end{cases}$$

Let $m = n + 1$ and conclude. Using the above formula we get $B_1 = -1/2$, $B_2 = 1/6$, $B_3 = 0$ and $B_4 = -1/30$. To show that $B_n = 0$ if n is odd $\neq 1$ we use Exercise 2. Let

$$f(z) = \frac{z}{e^z - 1} + \frac{z}{2}.$$

This eliminates the first term of the power series which defines the Bernoulli numbers, hence $f(z) = \sum_{n \neq 1} \frac{B_n}{n!} z^n$. Some straightforward computations show that

$$f(z) - f(-z) = z + z\frac{-1 + e^z + e^{-z} - 1}{1 - e^z - e^{-z} + 1} = 0.$$

Conclude using Exercise 2.

Exercise II.1.4. *Show that*

$$\frac{z}{2}\frac{e^{z/2}+e^{-z/2}}{e^{z/2}-e^{-z/2}} = \sum_{n=0}^{\infty}\frac{B_{2n}}{(2n)!}z^{2n}.$$

Replace z by $2\pi i z$ to show that

$$\pi z \cot \pi z = \sum_{n=0}^{\infty}(-1)^n\frac{(2\pi)^{2n}}{(2n)!}B_{2n}z^{2n}.$$

Solution. In Exercise 3 we proved that

$$f(z) = \frac{z}{e^z-1} + \frac{z}{2} = \sum_{n=0}^{\infty}\frac{B_{2n}}{(2n)!}z^{2n}.$$

We rewrite f as

$$f(z) = \frac{z}{2}\left(\frac{2}{e^z-1}+1\right) = \frac{z}{2}\frac{e^z+1}{e^z-1} = \frac{z}{2}\frac{e^{z/2}+e^{-z/2}}{e^{z/2}-e^{-z/2}}.$$

Combining these two results we obtain the desired formula. Replacing z by $2\pi i z$ in the left hand side of the above expression we obtain

$$\theta i \frac{e^{i\theta}+e^{-i\theta}}{e^{i\theta}-e^{-i\theta}} = \sum_{n=0}^{\infty}\frac{B_{2n}}{(2n)!}(2\pi i z)^{2n}$$

where $\theta = \pi z$. Euler's formulas imply $\cot\theta = i\frac{e^{i\theta}+e^{-i\theta}}{e^{i\theta}-e^{-i\theta}}$ and the desired formula drops out.

Exercise II.1.5. *Express the power series for* $\tan z$, $z/\sin z$, $z\cot z$, *in terms of Bernoulli numbers.*

Solution. Replacing πz by z in the last formula obtained in Exercise 4 we get

$$z\cot z = \sum_{n=0}^{\infty}(-1)^n\frac{2^{2n}}{(2n)!}B_{2n}z^{2n}.$$

The power series for $\tan z$ is obtained by using the above formula together with the identity

$$\tan z = \cot z - 2\cot 2z.$$

We find

$$z\tan z = \sum_{n=0}^{\infty}(-1)^n\frac{2^{2n}(1-2^{2n})}{(2n)!}B_{2n}z^{2n}.$$

The constant term is 0, so

$$\tan z = \sum_{n=0}^{\infty}(-1)^n\frac{2^{2n}(1-2^{2n})}{(2n)!}B_{2n}z^{2n-1}.$$

Finally we find the power series expansion of $z/\sin z$. The trigonometric formula we use is

$$\frac{2z}{\sin 2z} = 2z \cot z - 2z \cot 2z.$$

The above results then imply

$$\frac{2z}{\sin 2z} = \sum_{n=0}^{\infty} (-1)^n \frac{2^{2n}(2 - 2^{2n})}{(2n)!} B_{2n} z^{2n}.$$

Replacing $2z$ by z we find

$$\frac{z}{\sin z} = \sum_{n=0}^{\infty} (-1)^n \frac{2 - 2^{2n}}{(2n)!} B_{2n} z^{2n}.$$

Exercise II.1.6 (Difference Equations).   *Given complex numbers a_0, a_1, u_1, u_2 define a_n for $n \geq 2$ by*

$$a_n = u_1 a_{n-1} + u_2 a_{n-2}.$$

If we have a factorization

$$T^2 - u_1 T - u_2 = (T - \alpha)(T - \alpha'), \quad \text{and } \alpha \neq \alpha',$$

show that the numbers a_n are given by

$$a_n = A\alpha^n + B\alpha'^n$$

with suitable A, B. Find A, B in terms of $a_0, a_1, \alpha, \alpha'$. Consider the power series

$$F(T) = \sum_{n=0}^{\infty} a_n T^n.$$

Show that it represents a rational function, and give its partial fraction decomposition.

Solution. The existence of A and B is proved in the next exercise. Since $A + B = a_0$ and $A\alpha + B\alpha' = a_1$ we conclude that

$$A = \frac{a_1 - \alpha' a_0}{\alpha - \alpha'} \quad \text{and} \quad B = \frac{\alpha a_0 - a_1}{\alpha - \alpha'}.$$

If we consider the power series $F(T) = \sum_{n=0}^{\infty} a_n T^n$, we can write

$$F(T) = \sum_{n=0}^{\infty} A\alpha^n T^n + \sum_{n=0}^{\infty} B\alpha'^n T^n$$

$$= A \sum_{n=0}^{\infty} (\alpha T)^n + B \sum_{n=0}^{\infty} (\alpha' T)^n$$

$$= \frac{A}{1 - \alpha T} + \frac{B}{1 - \alpha' T}.$$

This gives us the desired representation of F as a rational function namely

$$F(T) = \frac{A + B - A\alpha'T - B\alpha T}{(1 - \alpha T)(1 - \alpha'T)}.$$

Exercise II.1.7. *More generally, let $a_0, \ldots, a_{r-1}$ be given complex numbers. Let $u_1, \ldots, u_r$ be complex numbers such that the polynomial*

$$P(T) = T^r - (u_1 T^{r-1} + \cdots + u_r)$$

has distinct roots $\alpha_1, \ldots, \alpha_r$. Define a_n for $n \geq r$ by

$$a_n = u_1 a_{n-1} + \cdots + u_r a_{n-r}.$$

Show that there exist numbers $A_1, \ldots, A_r$ such that for all n,

$$a_n = A_1 \alpha_1^n + \cdots + A_r \alpha_r^n.$$

Solution. The reader can find a solution to this exercise in the appendix of Lang's book. Here we give another argument.

Let (Σ) be the system

$$\begin{cases} a_0 = x_1 + \cdots + x_r \\ a_1 = x_1 \alpha_1 + \cdots + x_r \alpha_r \\ \quad \vdots \\ a_{r-1} = x_1 \alpha_1^{r-1} + \cdots + x_r \alpha_r^{r-1} \end{cases}$$

Since $\alpha_1, \ldots, \alpha_r$ are distinct,

$$\begin{vmatrix} 1 & \cdots & 1 \\ \alpha_1 & \cdots & \alpha_r \\ \vdots & \ddots & \vdots \\ \alpha_1^{r-1} & \cdots & \alpha_r^{r-1} \end{vmatrix} = \prod_{i \neq j}(\alpha_i - \alpha_j)$$

is nonzero (this is the Vandermonde determinant), so (Σ) has at least one solution, say $(A_1, \ldots, A_r)$. Now let S_n be the statement

$$a_n = \sum_{k=1}^{r} A_k \alpha_k^n.$$

By construction, $S_0, \ldots, S_{r-1}$ are true. Suppose S_n $(n \geq r - 1)$ is true for all $n \leq N$. Then

$$\begin{aligned} a_{N+1} &= u_1 a_N + \cdots + u_r a_{N+1-r} \\ &= u_1(A_1 \alpha_1^N + \cdots + A_r \alpha_r^N) + \cdots + u_r(A_1 \alpha_1^{N+1-r} + \cdots + A_r \alpha_r^{N+1-r}) \\ &= A_1 \alpha_1^{N+1-r}(u_1 \alpha_1^{r-1} + \cdots + u_r) + \cdots + A_r \alpha_r^{N+1-r}(u_1 \alpha_r^{r-1} + \cdots + u_r) \\ &= A_1 \alpha_1^{N+1} + \cdots + A_r \alpha_r^{N+1} \end{aligned}$$

where this last equality follows from the fact that $P(\alpha_i) = 0$ for all $1 \leq i \leq r$. So S_{N+1} is true, and by induction we conclude that S_n is true for all $n \geq 0$.

II.2   Convergent Power Series

Exercise II.2.1. *Let $|\alpha| < 1$. Express the sum of the geometric series*

$$\sum_{n=1}^{\infty} \alpha^n$$

in its usual simple form.

Solution. We simply use a geometric series

$$\sum_{n=1}^{m} \alpha^n = \frac{\alpha^{m+1} - \alpha}{\alpha - 1},$$

and since $|\alpha| < 1$ we have $\alpha^{m+1} \to 0$ as $m \to \infty$, so $\sum_{n=1}^{\infty} \alpha^n = \alpha/(1-\alpha)$ which is the expression we want.

Exercise II.2.2. *Let r be a real number, $0 \leq r < 1$. Show that the series*

$$\sum_{n=0}^{\infty} r^n e^{in\theta} \quad and \quad \sum_{n=-\infty}^{\infty} r^{|n|} e^{in\theta}$$

converge (θ is real). Express that series in simple terms using the usual formula for a geometric series.

Solution. We can use the comparison test because $|re^{i\theta}| = r$ and in Exercise 1 we showed that $\sum r^n$ converges. The expression of the series in simple terms is given by the following manipulation and the usual formula for the geometric series,

$$\sum_{n=0}^{\infty} r^n e^{in\theta} = \sum_{n=0}^{\infty} \left(re^{i\theta}\right)^n = \frac{1}{1 - re^{i\theta}}.$$

The second series converges for the exact same reasons the first series converged. To express the second series in a nice way we split the sum,

$$\sum_{n=-\infty}^{\infty} r^{|n|} e^{in\theta} = \sum_{n=-\infty}^{0} r^{|n|} e^{in\theta} + \sum_{n=1}^{\infty} r^n e^{in\theta} = \sum_{n=0}^{\infty} r^n e^{-in\theta} + \sum_{n=1}^{\infty} r^n e^{in\theta}$$

so that

$$\sum_{n=-\infty}^{\infty} r^{|n|} e^{in\theta} = \frac{1}{1 - re^{-i\theta}} + \frac{re^{i\theta}}{1 - re^{i\theta}}.$$

Exercise II.2.3. *Show that the usual power series for $\log(1+z)$ or $\log(1-z)$ from elementary calculus converges absolutely for $|z| < 1$.*

Solution. We have

$$\log(1 + z) = z - \frac{z^2}{2} + \cdots + \frac{(-1)^{n+1}}{n} z^n + \cdots = \sum_{n=1}^{\infty} \frac{(-1)^{n+1}}{n} z^n.$$

Taking the logarithm and using an elementary limit we find $\lim_{n \to \infty}(1/n)^{1/n} = 1$, so the radius of convergence of the above series is 1. Clearly, the proof and the result is the same for $\log(1-z)$.

Exercise II.2.4. *Determine the radius of convergence for the following power series.*

(a) $\sum n^n z^n$ (b) $\sum z^n/n^n$

(c) $\sum 2^n z^n$ (d) $\sum (\log n)^2 z^n$

(e) $\sum 2^{-n} z^n$ (f) $\sum n^2 z^n$

(g) $\sum \frac{n!}{n^n} z^n$ (h) $\sum \frac{(n!)^3}{(3n)!} z^n$

Solution. In all of this exercise we let a_n denote the coefficients of the power series which we are dealing with.

(a) Since $|a_n|^{1/n} = n$ it follows that the radius of convergence is 0.

(b) In this case we have $|a_n|^{1/n} = 1/n$ so the radius of convergence is ∞.

(c) We have $|a_n|^{1/n} = 2$ so the radius of convergence is $1/2$.

(d) For all large n, $1 \le \log n \le n$ holds so $1 \le |a_n|^{1/n} \le n^{2/n}$. Since $\lim_{n \to \infty} n^{2/n} = 1$ (take the logarithm and use an elementary limit) we conclude that the power series has radius of convergence equal to 1.

(e) Arguing like in (c) we find that the radius of convergence is equal to 2.

(f) The radius of convergence of the power series is 1 because of the limit $\lim_{n \to \infty} n^{2/n} = 1$.

(g) By Stirling's formula, $n! = n^n e^{-n} u_n$ with $\lim_{n \to \infty} u_n^{1/n} = 1$. Then we see that

$$|a_n|^{1/n} = \left(\frac{n^n e^{-n} u_n}{n^n} \right)^{1/n} = e^{-1} u_n^{1/n}$$

which implies that the radius of convergence is e.

(h) Using the notation of (g) we get

$$|a_n|^{1/n} = \left(\frac{n^{3n} e^{-3n} u_n^3}{(3n)^{3n} e^{-3n} u_{3n}} \right)^{1/n} \to \frac{1}{3^3}$$

so the radius of convergence of the power series is 27.

Exercise II.2.5. *Let $f(z) = \sum a_n z^n$ have radius of convergence $r > 0$. Show that the following series have the same radius of convergence:*

(a) $\sum n a_n z^n$ (b) $\sum n^2 a_n z^n$

(c) $\sum n^d a_n z^n$ *for any positive integer d*   (d) $\sum_{n \ge 1} n a_n z^{n-1}$

Solution. We are given that $\sum a_n z^n$ has a strictly positive radius of convergence so it is sufficient to investigate the limit of the term next to a_n. We show that in all four cases, the limit is 1. We do (c) first. We have

$$\log(n^{d/n}) = d \frac{\log n}{n} \to 0$$

as $n \to \infty$, so $\lim_{n \to \infty} n^{d/n} = 1$. Let $d = 1$ or $d = 2$ to get (a) and (b). For (d) we want to find the radius of convergence of the power series $\sum_{n=0}^{\infty} (n+1) a_{n+1} z^n$.

We can write

$$((n + 1)|a_{n+1}|)^{1/n} = \left((n + 1)^{1/(n+1)}|a_{n+1}|^{1/(n+1)}\right)^{(n+1)/n},$$

so the desired result follows.

Exercise II.2.6. *Give an example of power series whose radius of convergence is 1, and such that the corresponding function is continuous on the closed unit disc. [Hint: Try $\sum z^n/n^2$.]*

Solution. The power series $\sum z^n/n^2$ has a radius of convergence equal to 1 because

$$\lim_{n \to \infty} \left(\frac{1}{n^2}\right)^{1/n} = \lim_{n \to \infty} \left(\frac{1}{n^{1/n}}\right)^2 = 1.$$

To show that $\sum 1/n^2 < \infty$ we can either use the integral test with $h(x) = 1/x^2$ or use the fact that $1/n^2 < 1/n(n-1)$ and that the partial sums of $\sum 1/n(n-1)$ are telescopic. Let $f_n(z) = z^n/n^2$ and let $\overline{D}$ be the closed unit disc. Then on $\overline{D}$ we have

$$\|f_n\| \leq 1/n^2$$

for all n, where $\|\cdot\|$ denotes the sup norm on $\overline{D}$. So the series $\sum f_n$ converges uniformly on $\overline{D}$. Since the partial sums are continuous and the convergence is uniform, we conclude that the limit function, namely the power series $\sum z^n/n^2$, is continuous on $\overline{D}$.

Exercise II.2.7. *Let a, b be two complex numbers, and assume that b is not equal to any integer ≤ 0. Show that the radius of convergence of the series*

$$\sum \frac{a(a+1)\cdots(a+n)}{b(b+1)\cdots(b+n)} z^n$$

is at least 1. Show that this radius can be ∞ in some cases.

Solution. First, note that if a is equal to some negative integer, the series has only finitely many terms, and in this case the radius of convergence is ∞. Suppose a is never equal to some negative integer. Let

$$c_n = \frac{a(a+1)\cdots(a+n)}{b(b+1)\cdots(b+n)}$$

which is never equal to 0. Then

$$\left|\frac{c_{n+1}}{c_n}\right| = \frac{|a+n+1|}{|b+n+1|} = \frac{\left|\frac{a}{n+1} + 1\right|}{\left|\frac{b}{n+1} + 1\right|}$$

and the above ratio converges to 1 as $n \to \infty$. This concludes the proof.

Exercise II.2.8. *Let $\{a_n\}$ be a decreasing sequence of positive numbers approaching 0. Prove that the series $\sum a_n z^n$ is uniformly convergent on the domain of z such that*

$$|z| \leq 1 \text{ and } |z - 1| \geq \delta,$$

where $\delta > 0$. [Hint: For this problem and the next, use summation by parts.]

Solution. Let $T_n(z) = \sum_{k=0}^n a_k z^k$ and $S_n(z) = \sum_{k=0}^n z^k$. The summation by parts formula gives

$$T_n(z) = a_n S_n(z) - \sum_{k=0}^{n-1} S_k(z)(a_{k+1} - a_k)$$

hence if $n > m$ some straightforward computations show that

$$T_n(z) - T_m(z) = a_n(S_n(z) - S_m(z)) + \sum_{k=m+1}^{n-1} (S_k(z) - S_m(z))(a_k - a_{k+1})$$

Summing a geometric series we find $S_n(z) = (z^{n+1} - 1)/(z - 1)$, so using the assumption that $|z| \leq 1$ and $|z - 1| \geq \delta$ we get the uniform bound $|S_n(z)| \leq 2/\delta$ for all n. Therefore $|S_n(z) - S_m(z)| \leq 4/\delta$ for all m and n. Putting absolute values in the above displayed equation and using the triangle inequality and the fact that $\{a_n\}$ is positive and decreasing we get

$$|T_n(z) - T_m(z)| \leq a_n \frac{4}{\delta} + \frac{4}{\delta} \sum_{k=m+1}^{n-1} (a_k - a_{k+1})$$

$$= a_n \frac{4}{\delta} + \frac{4}{\delta}(a_{m+1} - a_n)$$

$$= \frac{4}{\delta} a_{m+1}.$$

Since $a_m \to 0$ as $m \to \infty$ we conclude that the series $\sum a_n z^n$ is uniformly convergent in the domain $|z| \leq 1$ and $|z - 1| \geq \delta$ of the complex plane.

Exercise II.2.9 (Abel's Theorem). *Let $\sum_{n=0}^{\infty} a_n z^n$ be a power series with radius of convergence ≥ 1. Assume that the series $\sum_{n=0}^{\infty} a_n$ converges. Let $0 \leq x < 1$. Prove that*

$$\lim_{x \to 1} \sum_{n=0}^{\infty} a_n x^n = \sum_{n=0}^{\infty} a_n.$$

Remark. *This result amounts to proving an interchange of limits. If*

$$s_n(x) = \sum_{k=1}^n a_k x^k,$$

then one wants to prove that

$$\lim_{n \to \infty} \lim_{x \to 1} s_n(x) = \lim_{x \to 1} \lim_{n \to \infty} s_n(x).$$

Solution. Let $f(x) = \sum_{k=1}^{\infty} a_k x^k$, $A = \sum_{k=1}^{\infty} a_k$ and $A_n = \sum_{k=1}^n a_k$. Consider the partial sums

$$s_n(x) = \sum_{k=1}^n a_k x^k.$$

We first prove that the sequence of partial sums $\{s_n(x)\}$ converges uniformly for $0 \le x \le 1$. For $m < n$, applying the summation by parts formula, we get

$$s_n(x) - s_m(x) = \sum_{k=m+1}^{n} x^k a_k = x^n(A_n - A_{m+1}) + \sum_{k=m+1}^{n-1} (A_k - A_{m+1})(x^k - x^{k+1}).$$

There exists N such that for $k, m \ge N$ we have $|A_k - A_{m+1}| \le \epsilon$. Hence for $0 \le x \le 1$ and $n, m \ge N$ we have

$$|s_n(x) - s_m(x)| \le \epsilon + \epsilon \sum_{k=m+1}^{n-1} (x^k - x^{k+1})$$

$$= \epsilon + \epsilon(x^{m+1} - x^n)$$

$$\le 3\epsilon$$

This proves the uniform convergence of $\{s_n(x)\}$.

Now given ϵ, pick N as above. Choose δ (depending on N) such that if $|x - 1| < \delta$, then

$$|S_N(x) - A_N| < \epsilon.$$

By combining the above results we find that

$$|f(x) - A| \le |f(x) - s_n(x)| + |s_n(x) - s_N(x)| + |s_N(x) - A_N| + |A_N - A|$$
$$\le |f(x) - s_n(x)| + 5\epsilon$$

for all $n \ge N$ and $|x - 1| < \delta$. For a given x, pick n so large (depending on x!) so that the first term is also $< \epsilon$, to conclude the proof. This argument is in the appendix of Lang's book.

Exercise II.2.10. *Let $\sum a_n z^n$ and $\sum b_n z^n$ be two power series, with radius of convergence r and s, respectively. What can you say about the radius of convergence of the series:*
(a) $\sum (a_n + b_n)z^n$ *(b)$\sum a_n b_n z^n$?*

Solution. (a) The triangle inequality implies

$$\sum_{n=0}^{m} |a_n + b_n| \, |z|^n \le \sum_{n=0}^{m} |a_n| \, |z|^n + \sum_{n=0}^{m} |b_n| \, |z|^n.$$

If $|z| < \min(r, s)$, then $\sum (a_n + b_n)z^n$ converges, so we can say that the radius of convergence of this last series is $\ge \min(r, s)$.

(b) Since $\limsup u_n v_n \le (\limsup u_n)(\limsup v_n)$ it follows that the radius of convergence of the series $\sum a_n b_n z^n$ is $\ge rs$.

Exercise II.2.11. *Let α, β be complex numbers with $|\alpha| < |\beta|$. Let*

$$f(z) = \sum (3\alpha^n - 5\beta^n)z^n.$$

Determine the radius of convergence of $f(z)$.

Solution. By Exercise 10 we see that the radius of convergence of the series is at least $1/|\beta|$. We contend that this radius of convergence is exactly $1/|\beta|$. Let $a_n = 3\alpha^n - 5\beta^n$. Then

$$|a_n| = 5|\beta|^n \left| \frac{3}{5}\left(\frac{\alpha}{\beta}\right)^n - 1 \right|$$

and since $|\alpha/\beta| < 1$ we have $\lim |a_n|^{1/n} = |\beta|$ thereby proving our contention.

Exercise II.2.12. *Let $\{a_n\}$ be the sequence of real numbers defined by the conditions:*

$$a_0 = 1, \quad a_1 = 2, \quad and \quad a_n = a_{n-1} + a_{n-2} \quad for \ n \geq 2.$$

Determine the radius of convergence of the power series

$$\sum_{n=0}^{\infty} a_n z^n.$$

[Hint: What is the general solution of a difference equation? Cf. Exercise 6 of §1.]

Solution. This exercise is a special case of Exercise 13.

Exercise II.2.13. *More generally, let u_1, u_2 be complex numbers such that the polynomial*

$$P(T) = T^2 - u_1 T - u_2 = (T - \alpha_1)(T - \alpha_2)$$

has two distinct roots with $|\alpha_1| < |\alpha_2|$. Let a_0, a_1 be given, and let

$$a_n = u_1 a_{n-1} + u_2 a_{n-2} \quad for \ n \geq 2.$$

What is the radius of convergence of the series $\sum a_n T^n$?

Solution. We solve the difference equation as was done in Exercise 6, §1 of this Chapter, so that we find

$$a_n = A\alpha_1^n + B\alpha_2^n$$

where A and B are complex numbers determined by a_0, a_1, α_1 and α_2. Then

$$|a_n| = B|\alpha_2|^n \left| \frac{A}{B}\left(\frac{\alpha_1}{\alpha_2}\right)^n + 1 \right|.$$

But $|\alpha_1/\alpha_2|^n \to 0$ so $|a_n|^{1/n} \to |\alpha_2|$. Therefore the radius of convergence of the series $\sum a_n z^n$ is $1/|\alpha_2|$.

II.3   Relations Between Formal and Convergent Series

Exercise II.3.1. *(a) Use the above definition of $\log z$ for $|z - 1| < 1$ to prove that $\exp \log z = z$. [Hint: What are the values on the left when $z = x$ is real?]*

(b) *Let* $z_0 \neq 0$. *Let* α *be a complex number such that* $\exp(\alpha) = z_0$. *For* $|z - z_0| < |z_0|$ *define*

$$\log z = f\left(\frac{z}{z_0} - 1\right) + \alpha.$$

Prove that $\exp \log z = z$ *for* $|z - z_0| < |z_0|$.

Solution. (a) For x real we have the usual series

$$e^x = \sum_{n=0}^{\infty} \frac{x^n}{n!}$$

$$\log x = \sum_{n=1}^{\infty} \frac{(-1)^{n-1}}{n} (x-1)^n \quad \text{for } 0 < x < 2.$$

We know that $\exp \log x = x$ for all $x > 0$, so if z is real and $|z - 1| < 1$ then

$$\exp \log z = z.$$

Combined with a translation to the origin, the uniqueness theorem (Theorem 3.2) implies the desired result.

(b) The given inequality implies that $|z/z_0 - 1| < 1$, so by (a)

$$\exp \log z = \left(\exp f\left(\frac{z}{z_0} - 1\right)\right) \cdot (\exp \alpha) = \frac{z}{z_0} z_0 = z.$$

Note that we have used the fact that $\exp(z_1 + z_2) = (\exp z_1)(\exp z_2)$ which can be easily proved using the uniqueness theorem twice.

Exercise II.3.2. (a) *Let* $\exp(T) = \sum_{n=0}^{\infty} T^n/n!$ *and*

$$\log(1 + T) = \sum_{k=1}^{\infty} (-1)^{k-1} T^k/k,$$

show that

$$\exp \log(1 + T) = 1 + T \text{ and } \log \exp(T) = T.$$

(b) *Let* $h_1(T)$ *and* $h_2(T)$ *be formal power series with* 0 *constant terms. Prove that* $\log((1 + h_1(T))(1 + h_2(T))) = \log(1 + h_1(T)) + \log(1 + h_2(T))$.
(c) *For complex numbers* α, β *show that* $\log(1 + T)^\alpha = \alpha \log(1 + T)$ *and*

$$(1 + T)^\alpha (1 + T)^\beta = (1 + T)^{\alpha + \beta}.$$

Solution. (a) For all real numbers $x > -1$ we know that

$$\exp \log(1 + x) = 1 + x$$

so by the uniqueness theorem we have

$$\exp \log(1 + T) = 1 + T.$$

Since $\log \exp x = x$ whenever x is real, we conclude that

$$\log \exp T = T.$$

(b) To prove this assertion, we first note that substituting power series in part (a) we find that $\exp \log(1 + h(T)) = 1 + h(T)$ for all power series $h(T)$ with 0 constant term. Therefore

$$
\begin{aligned}
\exp \log((1 + h_1(T))(1 + h_2(T))) &= \exp \log(1 + h_1(T) + h_2(T) + h_1(T)h_2(T)) \\
&= 1 + h_1(T) + h_2(T) + h_1(T)h_2(T) \\
&= (1 + h_1(T))(1 + h_2(T)) \\
&= \exp \log(1 + h_1(T)) \exp \log(1 + h_2(T)) \\
&= \exp(\log(1 + h_1(T)) + \log(1 + h_2(T)))
\end{aligned}
$$

So $\log((1 + h_1(T))(1 + h_2(T)))$ and $\log(1 + h_1(T)) + \log(1 + h_2(T))$ differ by a constant multiple of $2\pi i$. Evaluating at 0 we find that this constant is 0.

(c) If α is real, then for all real T we have the identity $\log(1 + T)^\alpha = \alpha \log(1 + T)$, so when α is real, we have equality of the two power series in T. Now $\log(1 + T)^\alpha$ is a power series with real coefficients which we may consider polynomials in α, namely $\log(1 + T)^\alpha = \sum a_n(\alpha)T^n$ where $a_n(\alpha)$ are real polynomials in α. Similarly, $\alpha \log(1 + T)$ is also a power series in T whose coefficients are real polynomials in the variable α, say, $\alpha \log(1 + T) = \sum b_n(\alpha)T^n$. Then, since $a_n(\alpha) = b_n(\alpha)$ for all real α, this equations also holds true for all complex α. Hence $\log(1 + T)^\alpha = \alpha \log(1 + T)$ for all complex α.

Since $(1 + T)^\alpha = 1 + h_\alpha(T)$ and $(1 + T)^\beta = 1 + h_\beta(T)$ where $h_\alpha(T)$ and $h_\beta(T)$ are power series with 0 constant term, we get from (b)

$$
\log((1 + T)^\alpha (1 + T)^\beta) = \log(1 + T)^\alpha + \log(1 + T)^\beta.
$$

Since $\log(1 + T)^\alpha = \alpha \log(1 + T)$ we find that

$$
\begin{aligned}
\log((1 + T)^\alpha (1 + T)^\beta) &= \alpha \log(1 + T) + \beta \log(1 + T) \\
&= (\alpha + \beta) \log(1 + T) \\
&= \log(1 + T)^{\alpha + \beta}.
\end{aligned}
$$

Exponentiating gives the desired result.

Exercise II.3.3. *Prove that for all complex z we have*

$$
\cos z = \frac{e^{iz} + e^{-iz}}{2} \quad and \quad \sin z = \frac{e^{iz} - e^{-iz}}{2}.
$$

Solution. We use the power series expansion of the exponential:

$$
e^{iz} = \sum_{n=0}^{\infty} i^n \frac{z^n}{n!} \quad and \quad e^{-iz} = \sum_{n=0}^{\infty} (-1)^n i^n \frac{z^n}{n!}.
$$

Adding these two series we obtain

$$
e^{iz} + e^{-iz} = \sum_{n \text{ even}} 2i^n \frac{z^n}{n!} = 2 \sum_{n=0}^{\infty} i^{2n} \frac{z^{2n}}{(2n)!} = 2 \cos z.
$$

Use the same argument to prove the formula for the sine.

Exercise II.3.4. *Show that the only complex numbers z such that $\sin z = 0$ are $z = k\pi$, where k is an integer. State and prove a similar statement for $\cos z$.*

Solution. By the previous exercise, the equation $\sin z = 0$ is equivalent to $e^{iz} = e^{-iz}$ or simply $e^{2iz} = 1$. Writing $z = x + iy$ where x and y are real we find that the equation $\sin z = 0$ is equivalent to $e^{2ix} e^{-2y} = 1$. Since e^{2ix} has absolute value 1 it follows that $e^{-2y} = 1$ so $y = 0$. Therefore the real part of z must verify $e^{2ix} = 1$, so $x = k\pi$ with $k \in \mathbf{Z}$. For the cosine the exact same argument shows that $\cos z = 0$ if and only if $z = k\pi/2$ where k is an integer.

Exercise II.3.5. *Find the power series expansion of $f(z) = 1/(z+1)(z+2)$, and find the radius of convergence.*

Solution. We use partial fraction decomposition to modify the expression of f, namely

$$\frac{1}{(z+1)(z+2)} = \frac{1}{z+1} - \frac{1}{z+2} = \frac{1}{1-(-z)} - \frac{1}{2}\frac{1}{1-(-z/2)}.$$

The formula for the sum of a geometric series implies that for $|z| < 1$ we get

$$f(z) = \sum_{n=0}^{\infty}(-z)^n - \frac{1}{2}\sum_{n=0}^{\infty}(-z/2)^n$$

$$= \sum_{n=0}^{\infty}\left[(-1)^n - \frac{(-1)^n}{2^{n+1}}\right]z^n$$

$$= \sum_{n=0}^{\infty}(-1)^n\left(1 - \frac{1}{2^{n+1}}\right)z^n.$$

Since $\limsup\left(1 - \frac{1}{2^{n+1}}\right)^{1/n} = 1$, the radius of convergence of the power series is 1.

Exercise II.3.6. *The **Legendre polynomials** can de defined as the coefficients $P_n(\alpha)$ of the series expansion of*

$$f(z) = \frac{1}{(1 - 2\alpha z + z^2)^{1/2}}$$

$$= 1 + P_1(\alpha)z + P_2(\alpha)z^2 + \cdots + P_n(\alpha)z^n + \cdots .$$

Calculate the first four Legendre polynomials.

Solution. Squaring both sides we get

$$\frac{1}{1 - 2\alpha z + z^2} = \left(\sum_{n=0}^{\infty}P_n(\alpha)z^n\right)^2 = \sum_{n=0}^{\infty}d_n(\alpha)z^n$$

where $P_0(\alpha) = 1$ and $d_n(\alpha) = \sum_{k=0}^{n}P_n(\alpha)P_{n-k}(\alpha)$. Hence

$$1 = (1 - 2\alpha z + z^2)\left(\sum_{n=0}^{\infty}d_n(\alpha)z^n\right)$$

$$= \sum_{n=0}^{\infty} d_n(\alpha)z^n - 2\alpha \sum_{n=0}^{\infty} d_n(\alpha)z^{n+1} + \sum_{n=0}^{\infty} d_n(\alpha)z^{n+2}$$

$$= d_0(\alpha) + (d_1(\alpha) - 2\alpha d_0(\alpha)) + \sum_{n=2}^{\infty}(d_n(\alpha) - 2\alpha d_{n-1}(\alpha) + d_n(\alpha))z^n.$$

We get the recursive formulas

$$\begin{cases} d_0(\alpha) = 1 \\ d_1(\alpha) - 2\alpha d_0(\alpha) = 0 \\ d_n(\alpha) - 2\alpha d_{n-1}(\alpha) + d_n(\alpha) = 0 \end{cases}$$

which allow us to find the desired Legendre polynomials. We find $P_1(\alpha) = \alpha$, $P_2(\alpha) = \frac{1}{2}(3\alpha^2 - 1)$, $P_3(\alpha) = \frac{1}{2}(5\alpha^3 - 3\alpha)$ and for the fourth polynomial we find $P_4(\alpha) = \frac{1}{8}(35\alpha^4 - 30\alpha^2 + 3)$.

Note. It can be shown that $P_n(\alpha) = \frac{1}{2^n n!} D^n \left[(\alpha^2 - 1)^n\right]$.

II.4   Analytic Functions

Exercise II.4.1. *Find the terms of order ≤ 3 in the power series expansion of the function $f(z) = z^2/(z - 2)$ at $z = 1$.*

Solution. To find the expansion we must modify the expression on f. We write

$$z^2 = [(z - 1) + 1]^2 = (z - 1)^2 + 2(z - 1) + 1$$

and

$$z - 2 = -(1 - (z - 1)),$$

so that we find

$$f(z) = [(z - 1)^2 + 2(z - 1) + 1]\left(-\sum_{k=0}^{\infty}(z - 1)^k\right).$$

Therefore the beginning of the power series expansion of f is

$$-1 - (1 + 2)(z - 1) - (1 + 2 + 1)(z - 1)^2 - (1 + 2 + 1)(z - 1)^3.$$

Exercise II.4.2. *Find the terms of order ≤ 3 in the power series expansion of the function $f(z) = (z - 2)/(z + 3)(z + 2)$ at $z = 1$.*

Solution. We use partial fraction decomposition to get

$$f(z) = \frac{5}{z + 3} - \frac{4}{z + 2}.$$

But

$$\frac{5}{z + 3} = \frac{5}{4} \frac{1}{1 - \frac{-1}{4}(z - 1)}$$

$$= \frac{5}{4}\left(1 - \frac{1}{4}(z-1) + \frac{1}{4^2}(z-1)^2 - \frac{1}{4^3}(z-1)^3 + \cdots\right)$$

and the same method shows that

$$\frac{4}{z+2} = \frac{4}{3}\left(1 - \frac{1}{3}(z-1) + \frac{1}{3^2}(z-1)^2 - \frac{1}{3^3}(z-1)^3 + \cdots\right).$$

Therefore we find that the beginning of the power series of f at 1 is

$$\left(\frac{5}{4} - \frac{4}{3}\right) - \left(\frac{5}{4^2} - \frac{4}{3^2}\right)(z-1) + \left(\frac{5}{4^3} - \frac{4}{3^3}\right)(z-1)^2 - \left(\frac{5}{4^4} - \frac{4}{3^4}\right)(z-1)^3.$$

II.5   Differentiation of Power Series

In Exercises 1 through 5, also determine the radius of convergence of the given series.

Exercise II.5.1. *Let*

$$f(z) = \sum \frac{z^{2n}}{(2n)!}$$

Prove that $f''(z) = f(z)$.

Solution. The ratio test implies at once that the radius of convergence is ∞. Let $g_n(z) = z^{2n}/(2n)!$. Then

$$g_n'(z) = \frac{2n}{(2n)!}z^{2n-1} \quad \text{and} \quad g_n''(z) = \frac{2n(2n-1)}{(2n)!}z^{2n-2} = g_{n-1}(z),$$

which implies that $f(z) = f''(z)$.

Exercise II.5.2. *Let*

$$f(z) = \sum_{n=0}^{\infty} \frac{z^{2n}}{(n!)^2}.$$

Prove that

$$z^2 f''(z) + z f'(z) = 4z^2 f(z).$$

Solution. By the ratio test we see that the radius of convergence of the series is ∞. Differentiating we find

$$f'(z) = \sum_{n=1}^{\infty} \frac{2n}{(n!)^2}z^{2n-1} \quad \text{and} \quad f''(z) = \sum_{n=1}^{\infty} \frac{2n(2n-1)}{(n!)^2}z^{2n-2}$$

so that

$$zf'(z) = \sum_{n=1}^{\infty} \frac{2n}{(n!)^2}z^{2n}$$

and

$$z^2 f''(z) = \sum_{n=1}^{\infty} \frac{2n(2n-1)}{(n!)^2} z^{2n} = 4 \sum_{n=1}^{\infty} \frac{n^2}{(n!)^2} z^{2n} - z f'(z)$$

Therefore

$$z^2 f''(z) + z f'(z) = 4 \sum_{n=1}^{\infty} \frac{n^2}{(n!)^2} z^{2n} = 4 \sum_{n=1}^{\infty} \frac{1}{((n-1)!)^2} z^{2n} = 4z^2 f(z)$$

as was to be shown.

Exercise II.5.3. *Let*

$$f(z) = z - \frac{z^3}{3} + \frac{z^5}{5} - \frac{z^7}{7} + \cdots.$$

Show that $f'(z) = 1/(z^2 + 1)$.

Solution. Since $\limsup(1/n)^{1/n} = 1$ the radius of convergence is 1. Differentiating term by term we get

$$f'(z) = 1 - z^2 + z^4 - \cdots = \sum_{n=0}^{\infty} (-z^2)^n = \frac{1}{1 - (-z^2)} = \frac{1}{1 + z^2}.$$

Exercise II.5.4. *Let*

$$J(z) = \sum_{n=0}^{\infty} \frac{(-1)^n}{(n!)^2} \left(\frac{z}{2}\right)^{2n}.$$

Prove that

$$z^2 J''(z) + z J'(z) + z^2 J(z) = 0.$$

Solution. The ratio test implies that the radius of convergence is ∞. Let $S(z) = z^2 J''(z) + z J'(z) + z^2 J(z)$. Then differentiating term by term, some elementary manipulations show that

$$S(z) = \sum_{n=0}^{\infty} (z^2 + 2n + 2n(2n-1)) \frac{(-1)^n}{(n!)^2 2^{2n}} z^{2n}$$

$$= \sum_{n=0}^{\infty} (z^2 + 4n^2) \frac{(-1)^n}{(n!)^2 2^{2n}} z^{2n}.$$

However

$$(z^2 + 4n^2) \frac{(-1)^n}{(n!)^2 2^{2n}} z^{2n} = \frac{(-1)^n}{(n!)^2 2^{2n}} z^{2(n+1)} - \frac{(-1)^{n-1}}{((n-1)!)^2 2^{2(n-1)}} z^{2n}$$

so we have a telescopic sum, which implies that $S(z) = 0$, as was to be shown.

Exercise II.5.5. *For any positive integer k, let*

$$J_k(z) = \sum_{n=0}^{\infty} \frac{(-1)^n}{n!(n+k)!} \left(\frac{z}{2}\right)^{2n+k}.$$

Prove that

$$z^2 J_k''(z) + z J_k'(z) + (z^2 - k^2) J_k(z) = 0.$$

Solution. The ratio test implies that the radius of convergence is ∞. Let

$$a_n = \frac{(-1)^n}{2^{2n+k} n!(n+k)!}$$

Differentiating term by term we find that the coefficient of z^{2n+k} in the sum $z^2 J_k''(z) + z J_k'(z) + (z^2 - k^2) J(z)$ is

$$= a_n(2n+k)(2n+k-1) + a_n(2n+k) + a_{n-1} - k^2 a_n$$
$$= (4kn + 4n^2)a_n + a_{n-1}$$

But $a_{n-1}/a_n = -4n(n+k)$ so $z^2 J_k''(z) + z J_k'(z) + (z^2 - k^2) J_k(z) = 0$ as was to be shown.

Exercise II.5.6. *(a) For $|z-1| < 1$, show that the derivative of the function*

$$\log z = \log(1 + (z-1)) = \sum_{n=1}^{\infty} (-1)^{n-1} \frac{(z-1)^n}{n}$$

is $1/z$.
(b) Let $z_0 \neq 0$. For $|z - z_0| < 1$, define $f(z) = \sum (-1)^{n-1}((z-z_0)/z_0)^n/n$. Show that $f'(z) = 1/z$.

Solution. (a) Differentiating term by term we find that

$$(\log z)' = \sum_{n=1}^{\infty} (-1)^{n-1}(z-1)^{n-1} = \frac{1}{1-(-(z-1))} = \frac{1}{z}.$$

(b) Differentiating term by term we find that

$$f'(z) = \sum_{n=1}^{\infty} \frac{(-1)^{n-1}}{z_0^n}(z-z_0)^{n-1} = \sum_{n=1}^{\infty} \frac{(-1)^{n-1}}{z_0}\left(\frac{z}{z_0}-1\right)^{n-1}.$$
$$= \frac{1}{z_0}\frac{1}{1-(z/z_0-1)} = \frac{1}{z}.$$

II.6 The Inverse and Open Mapping Theorems

Determine which of the following functions are local analytic isomorphism at the given point. Give the reason for your answer.

Exercise II.6.1. $f(z) = e^z$ at $z = 0$.

Solution. By definition we have $f(z) = \sum z^n/n!$ and $f'(z) = f(z)$, so $f'(0) = 1 \neq 0$. Therefore f is a local analytic isomorphism at 0.

Exercise II.6.2. $f(z) = \sin(z^2)$ at $z = 0$.

Solution. For all z near zero we have $\sin(z^2) = \sin((-z)^2)$ so there does not exist and open ball around 0 such that the given function is an analytic isomorphism in this ball.

Exercise II.6.3. $f(z) = (z-1)/(z-2)$ *at* $z = 1$.

Solution. We write $z - 2 = -(1 - (z-1))$ so that for all z near 1 we have

$$f(z) = -(z-1)\left[1 + (z-1) + (z-1)^2 + \cdots\right]$$

so that $f'(1) = -1$. This proves that f is a local analytic isomorphism at $z = 1$.

Exercise II.6.4. $f(z) = (\sin z)^2$ *at* $z = 0$.

Solution. For all z near 0 we have $(\sin z)^2 = (\sin -z)^2$ and therefore f cannot be a local isomorphism at 0.

Exercise II.6.5. $f(z) = \cos z$ *at* $z = \pi$.

Solution. For all real x near 0 we have $\cos(\pi - x) = \cos(\pi + x)$ so f is not a local analytic isomorphism at 0.

Exercise II.6.6 (Linear Differential Equations). *Prove:*

Theorem. *Let* $a_0(z), \ldots, a_k(z)$ *be analytic functions in a neighborhood of 0. Assume that* $a_0(0) \neq 0$. *Given numbers* $c_0, \ldots, c_{k-1}$, *there exists a unique analytic function* f *at 0 such that*

$$D^n f(0) = c_n \quad for\ n = 0, \ldots, k-1$$

and such that

$$a_0(z)D^k f(z) + a_1(z)D^{k-1} f(z) + \cdots + a_k(z)f(z) = 0.$$

[Hint: First you may assume $a_0(z) = 1$ *(why?). Then solve for* f *by a formal power series. Then prove that this formal series converges.]*

Solution. The solution of this exercise is in the appendix of Lang's book but for the sake of completeness we will repeat the argument here. Since $a(0) \neq 0$ we must have $a(z) \neq 0$ in a neighborhood of 0 which shows after dividing through by a_0 that we may assume $a_0 = 1$. We change notation a little to be in accordance with the appendix of the book. In fact, we prove the following

Theorem. Let p be an integer ≥ 2. Let $g_0, \ldots, g_{p-1}$ be power series with complex coefficients. Let $a_0, \ldots, a_{p-1}$ be given complex numbers. Then there exists a unique power series $f(T) = \sum a_n T^n$ such that

$$D^p f = g_{p-1} D^{p-1} f + \cdots + g_0 f.$$

If $g_0, \ldots, g_{p-1}$ converge in a neighborhood of the origin, then so does f.

The coefficient a_n $(n \geq p)$ of f will be determined inductively and uniquely. Then we prove that the power series $\sum a_n z^n$ converges in a neighborhood of the origin. Proceeding formally we see that

$$D^p f(T) = \sum_{n=p}^{\infty} n(n-1)\cdots(n-p+1)a_n T^{n-p},$$

and therefore putting $m = n - p$, we get

$$D^p f(T) = \sum_{m=0}^{\infty} (m+p)(m+p-1)\cdots(m+1)a_{m+p}T^m.$$

Similarly, for every positive integer s with $0 \le s \le p-1$ we have

$$D^s f(T) = \sum_{k=0}^{\infty} (k+s)(k+s-1)\cdots(k+1)a_{k+s}T^k.$$

It will be useful to use the notation $[k, s] = (k+s)(k+s-1)\cdots(k+1)$ for $s \ge 1$ and $[k, s] = 1$ if $s = 0$.

Next we write down the power series for each g_s, say

$$g_s = \sum_{j=0}^{\infty} b_{s,j} T^j.$$

Then

$$g_s D^s f(T) = \sum c_{s,m} T^m \quad \text{where } c_{s,m} = \sum_{k+j=m} [k, s]a_{k+s}b_{s,j}. \tag{1}$$

Hence once we are given $a_0, \ldots, a_{p-1}$ we can solve inductively for a_m in terms of $a_0, \ldots, a_{m-1}$ and the coefficients of $g_1, \ldots, g_s$ by the formula

$$a_{m+p} = \frac{c_{0,m} + \cdots + c_{p-1,m}}{[m, p]} = \sum_{s=0}^{p-1} \sum_{k+j=m} \frac{[k, s]}{[m, p]} a_{k+s}b_{s,j} \tag{2}$$

which determines a_{m+p} uniquely in terms of $a_0, \ldots, a_{m+p-1}, b_{s,1}, \ldots, b_{s,m}$. Hence we have proved that there is a unique power series satisfying the differential equation.

Assuming that the power series $g_0, \ldots, g_{p-1}$ converge, we must now prove that $f(T)$ converges. We select a positive number K sufficiently large ≥ 2 and a positive number B such that

$$|a_0| \le K, \ldots, |a_{p-1}| \le K$$

and for all $s = 0, \ldots, p-1$ and all j we have

$$|b_{s,j}| \le K B^j.$$

We prove by induction that for $m \ge 0$ we have

$$|a_{m+p}| \le 2^m p^{m+1} K^2 K^m B^m. \tag{3}$$

The standard m-th root test for convergence then shows that $f(T)$ converges.

We note that the expression (1) for $c_{s,m}$ and hence (2) for a_{m+p} have positive coefficients as polynomials in $a_0, a_1, \ldots$ and the coefficients $b_{s,j}$ of the power series

g_s. Hence to make our estimates, we may avoid writing down absolute values by replacing $b_{s,j}$ by KB^j, and we may replace $a_0, \ldots, a_{p-1}$ by K. Then all values a_{m+p} ($m \geq 0$) are positive and we want to show that they satisfy the bound in (3).

Observe first that for $0 \leq k \leq m$ and $1 \leq s \leq p-1$ we always have

$$\frac{[k, s]}{[m, p]} \leq 1.$$

Hence the fraction $[k, s]/[m, p]$ will be replaced by 1 in the following estimates.

Now first we estimate a_{m+p} with $m = 0$. Then $k + j = 0$, so $k = j = 0$ and

$$a_p \leq \sum_{s=0}^{p-1} \sum_{k+j=0} a_{k+s} b_{s,j} \leq pK^2$$

as desired. Suppose by induction that we have proved (3) for all integers ≥ 0 and $< n$. Then

$$a_{n+p} \leq \sum_{s=0}^{p-1} \sum_{k+j=n} a_{k+s} b_{s,j} \leq p \sum_{k+j=n} 2^{n-1} p^n K^2 K^{k-1} B^k K B^j$$

$$\leq 2^{n-1} p^{n+1} \sum_{k+j=n} K^3 K^{k-1} B^k B^j$$

$$\leq 2^{n-1} p^{n+1} K^2 \sum_{k=0}^{n} K^k B^n$$

$$\leq 2^n p^{n+1} K^{n+2} B^n$$

which is the desired estimate. We have used the elementary inequality

$$K^2 \sum_{k=0}^{n} K^k = K^2 \frac{K^{n+1} - 1}{K - 1} \leq \frac{K}{K - 1} K^{n+2} \leq 2K^{n+2},$$

which is trivial.

Exercise II.6.7 (Ordinary Differential Equations). *Prove:*

Theorem. *Let g be analytic at 0. There exists a unique analytic function f at 0 satisfying*

$$f(0) = 0 \quad and \quad f'(z) = g(f(z)).$$

[Hint: Again find a formal solution, and then prove that it converges.]

Solution. Again, the proof is in the appendix of Lang's book, but we repeat the argument for sake of completeness.

Let $g(T) = \sum b_k T^k$ and write $f(T)$ with unknown coefficients

$$f(T) = \sum_{m=1}^{\infty} a_m T^m.$$

Then $f'(T) = \sum m a_m T^{m-1} = \sum (n+1) a_{n+1} T^n$. The given differential equation has the form

$$\sum_{n=0}^{\infty} (n+1) a_{n+1} T^n = b_0 + b_1 f(T) + b_2 f(T)^2 + \cdots.$$

Equating the coefficient of T^n on both sides, we see that $a_1 = b_0$, and

$$(n+1) a_{n+1} = P_n(b_0, \ldots, b_n; a_1, \ldots, a_n)$$

where P_n is a polynomials with positive integer coefficients. In particular, starting with $a_1 = b_0$, we may then solve inductively for a_{n+1} in terms of $a_1, \ldots, a_n$ for $n \geq 1$. This proves the existence and uniqueness of the power series $f(T)$.

Assume next that $g(T)$ converges. We must prove that $f(T)$ converges. Let B_k be positive numbers such that $|b_k| \leq B_k$, and such that the power series $G(T) = \sum B_k T^k$ converges. Let $F(T)$ be the solution of the differential equation

$$F'(T) = G(F(T)),$$

and let $F(T) = \sum A_m T^m$, with $A_1 > 0$ and $|a_1| \leq A_1$. Then

$$(n+1) A_{n+1} = P_n(B_0, \ldots, B_n; A_1, \ldots, A_n),$$

with the same polynomial P_n. Hence $|a_{n+1}| \leq A_{n+1}$, and if $F(T)$ converges so does $f(T)$.

Since $g(T)$ converges, there exists positive numbers K, B such that

$$|b_k| \leq K B^k \quad \text{for all } k = 0, 1, \ldots.$$

We let $B_k = K B^k$. Then

$$G(T) = K \sum_{k=0}^{\infty} B^k T^k = \frac{K}{1 - BT},$$

and so it suffices to prove that the solution $F(T)$ of the differential equation

$$F'(T) = \frac{K}{1 - BT}$$

converges. This equation is equivalent with

$$F'(T) = K + B F(T) F'(T),$$

which we can integrate to give

$$F(T) = KT + B \frac{F(T)^2}{2}.$$

By the quadratic formula, we find

$$F(T) = \frac{1 - (1 - 2K BT)^{1/2}}{B} = KT + \cdots.$$

We then use the binomial expansion which we know converges. This concludes the proof.

III

Cauchy's Theorem, First Part

III.1   Holomorphic Functions on Connected Sets

Exercise III.1.1. *Prove Lemma 1.5, that is, prove*

Lemma. *Let S be a subset of an open set U. Then S is closed on U if and only if the complement of S in U, that is, $U - S$ is open. In particular, if S is both open and closed in U, then $U - S$ is also open and closed in U.*

Solution. Suppose S is closed, and let $w \in U - S$. Since U is open, $U - S$ is not empty. Then for some n the ball of radius $1/n$ centered at w is contained in U and this ball does not intersect S, for otherwise w is in the closure of S which contradicts the fact that S is closed and $w \in S$. Hence $U - S$ is open.

Conversely, suppose that the set $U - S$ is open, and let $z \in U$ and $z \in \overline{S}$, where $\overline{S}$ denotes the closure of S. If z is not in S then there exists a ball centered at z which is contained in $U - S$. So z is not adherent to S, a contradiction which ends the proof.

Exercise III.1.2. *Let U be a bounded open connected set, $\{f_n\}$ a sequence of continuous functions on the closure of U, analytic on U. Assume that $\{f_n\}$ converges uniformly on the boundary of U. Prove that $\{f_n\}$ converges uniformly on U.*

Solution. We use Cauchy's criterion for uniform convergence. Let $\|\cdot\|_{\partial U}$ and $\|\cdot\|_U$ denote the sup norm on the boundary of U and the sup norm on U respectively. Let $\epsilon > 0$ and choose N such that for all $n, m > N$ we have $\|f_n - f_m\|_{\partial U} < \epsilon$. By the maximum modulus principle, the function $f_n - f_m$ attains its maximum on the boundary of U, so we have $\|f_n - f_m\|_U < \epsilon$ for all $n, m > N$ which implies the uniform convergence of the sequence $\{f_n\}$ on U.

Exercise III.1.3. *Let $a_1, \ldots, a_n$ be points on the unit circle. Prove that there exists a point z on the unit circle so that the product of the distances from z to the a_j is at least 1. (You may use the maximum modulus principle.)*

Solution. Consider the map $f : \overline{D}(0, 1) \to \mathbf{C}$ given by

$$f(z) = (z - a_1) \cdots (z - a_n) = \prod_{k=1}^{n} (z - a_k).$$

The closed unit disc is compact and the function f is continuous on $\overline{D}(0, 1)$ and analytic in the open unit disc. We have $|f(0)| = 1$ and $|f(a_k)| = 0$ for all $k = 1, \ldots, n$ which implies that f is not constant. By the maximum modulus principle, f attains its maximum at a point z of the unit circle, so $|f(z)| \geq |f(0)| = 1$.

III.2   Integrals over Paths

Exercise III.2.1. *(a) Given an arbitrary point z_0, let C be a circle of radius $r > 0$ centered at z_0, oriented counterclockwise. Find the integral*

$$\int_C (z - z_0)^n dz$$

for all integers n, positive or negative.
(b) Suppose f has a power series expansion

$$f(z) = \sum_{k=-m}^{\infty} a_k (z - z_0)^k$$

which is absolutely convergent on a disc of radius $> R$ centered at z_0. Let C_R be the circle of radius R centered at z_0. Find the integral

$$\int_{C_R} f(z) dz.$$

Solution. (a) Suppose $n \neq -1$ and let $f(z) = (z - z_0)^n$. Then if

$$g(z) = \frac{1}{n+1}(z - z_0)^{n+1}$$

we see that $g' = f$ so f has a primitive and since C is closed we find that

$$\int_C (z - z_0)^n dz = 0.$$

If $n = -1$ we parametrize C by $\gamma(t) = z_0 + re^{it}$ with $t \in [0, 2\pi]$, and we get

$$\int_C \frac{dz}{z - z_0} = \int_0^{2\pi} \frac{\gamma'(t)}{\gamma(t) - z_0} dt = \int_0^{2\pi} \frac{rie^{it}}{re^{it}} dt = 2\pi i.$$

(b) Let $f_n(z) = \sum_{k=-m}^{n} a_k(z - z_0)^k$. The sequence $\{f_n\}$ converges uniformly on C_R hence

$$\lim_{n\to\infty} \int_{C_R} f_n = \int_{C_R} f$$

The results obtained in (a) imply that

$$\int_{C_R} f(z)dz = 2\pi i a_{-1}.$$

Exercise III.2.2. *Find the integral of $f(z) = e^z$ from -3 to 3 taken along a semicircle. Is this integral different from the integral taken over the line segment between the two points?*

Solution. Since $f' = f$, the function f is its own primitive. Therefore the integral is independent of the path and is equal to $f(3) - f(-3)$.

Exercise III.2.3. *Sketch the following curves with $0 \le t \le 1$.*
(a) $\gamma(t) = 1 + it$
(b) $\gamma(t) = e^{-\pi it}$
(c) $\gamma(t) = e^{\pi it}$
(d) $\gamma(t) = 1 + it + t^2$

Solution.

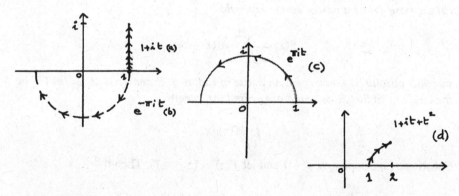

Exercise III.2.4. *Find the integral of each one of the following functions over each one of the curves in Exercise 3.*
(a) $f(z) = z^3$
(b) $f(z) = \bar{z}$
(c) $f(z) = 1/z$

Solution. (a) (i) $\frac{(1+i)^4}{4} - \frac{1}{4}$ (ii) 0 (iii) 0 (iv) $\frac{(2+i)^4}{4} - \frac{1}{4}$
(b) (i) $i + 1/2$ (ii) $-\pi i$ (iii) πi (iv) $2 + \frac{2i}{3}$.
(c) (i) $\log\sqrt{2} + i\frac{\pi}{4}$ (ii) $-\pi i$ (iii) πi (iv) $\log\sqrt{5} + i\arctan(1/2)$

Exercise III.2.5. *Find the integral*

$$\int_{\gamma} ze^{z^2} dz$$

(a) from the point i to the point $-i+2$, taken along the straight line segment, and
(b) from 0 to $1+i$ along the parabola $y = x^2$.

Solution. We see that the function g defined by $g(z) = \frac{1}{2}e^{z^2}$ is a primitive for ze^{z^2} so for (a) the integral is equal to

$$g(-i+2) - g(i) = \frac{1}{2}(e^{3-4i} - e^{-1})$$

and for (b) the integral is equal to

$$g(1+i) - g(0) = \frac{1}{2}(e^{2i} - 1).$$

Exercise III.2.6. *Find the integral*

$$\int_{\gamma} \sin z dz$$

from the origin to the point $1+i$, taken along the parabola

$$y = x^2.$$

Solution. The function g defined by $g(z) = -\cos z$ is a holomorphic primitive of $\sin z$ on $\mathbf{C}$, so

$$\int_{\gamma} \sin z dz = g(1+i) - g(0) = -(\cos(1+i) - 1)$$

Exercise III.2.7. *Let σ be a vertical segment, say parametrized by*

$$\sigma(t) = z_0 + itc, \quad -1 \le t \le 1,$$

where z_0 is a fixed complex number, and c is a fixed real number > 0. (Draw a picture.) Let $\alpha = z_0 + x$ and $\alpha' = z_0 - x$, where x is real positive. Find

$$\lim_{x \to 0} \int_{\sigma} \left(\frac{1}{z-\alpha} - \frac{1}{z-\alpha'} \right) dz.$$

(Draw the picture.) Warning: The answer is not 0!

Solution. The picture is the following:

Since $\sigma'(t) = ic$ we have

$$\int_\sigma \left(\frac{1}{z-\alpha} - \frac{1}{z-\alpha'} \right) dz = ic \int_{-1}^{1} \frac{1}{itc-x} - \frac{1}{itc+x} dt$$

Reducing to the same denominator we find that this last integral is equal to

$$= ic \int_{-1}^{1} \frac{2x}{-t^2 c^2 - x^2} dt = \frac{-2ic}{x} \int_{-1}^{1} \frac{1}{(tc/x)^2 + 1} dt$$

$$= -\frac{2ic}{x} \left[\frac{x}{c} \arctan(tc/x) \right]_{-1}^{1} = -4i \arctan(c/x).$$

The limit $\lim_{u \to \infty} \arctan(u) = \pi/2$ implies that

$$\lim_{x \to 0} \int_\sigma \left(\frac{1}{z-\alpha} - \frac{1}{z-\alpha'} \right) dz = -2\pi i.$$

Exercise III.2.8. *Let $x > 0$. Find the limit:*

$$\lim_{B \to \infty} \int_{-B}^{B} \left(\frac{1}{t+ix} - \frac{1}{t-ix} \right) dt.$$

Solution. We write

$$\frac{1}{t+ix} - \frac{1}{t-ix} = \frac{-2ix}{t^2 + x^2} = \frac{-2i}{x} \frac{1}{(t/x)^2 + 1}$$

so the integral in the limit is

$$= \frac{-2i}{x} [x \arctan(t/x)]_{-B}^{B} = -4i \arctan(B/x).$$

Whence

$$\lim_{B \to \infty} \int_{-B}^{B} \left(\frac{1}{t+ix} - \frac{1}{t-ix} \right) dt = -2\pi i.$$

Exercise III.2.9. *Let $\gamma : [a, b] \to \mathbf{C}$ be a curve. Define the **reverse** or **opposite** curve to be*

$$\gamma^- : [a, b] \to \mathbf{C}$$

such that $\gamma^-(t) = \gamma(a + b - t)$. Show that

$$\int_{\gamma^-} F = -\int_{\gamma} F.$$

Solution. The integral along the reverse path is

$$\int_{\gamma^-} F = \int_a^b -F(\gamma(a + b - t))\gamma'(a + b - t)dt.$$

Changing variables $u = a + b - t$ we obtain

$$\int_{\gamma^-} F = \int_b^a F(\gamma(u))\gamma'(u)du = -\int_{\gamma} F,$$

as was to be shown.

Exercise III.2.10. *Let $[a, b]$ and $[c, d]$ be two intervals (not reduced to a point). Show that there is a function $g(t) = rt + s$ such that g is strictly increasing, $g(a) = c$ and $g(b) = d$. Thus a curve can be parametrized by any given interval.*

Solution. By putting $r = \frac{d-c}{b-a}$ and $s = c - ra$ we get the desired function g.

Exercise III.2.11. *Let F be a continuous complex-valued function on the interval $[a, b]$. Prove that*

$$\left| \int_a^b F(t)dt \right| \leq \int_a^b |F(t)|dt.$$

[Hint: Let $P = [a = a_0, a_1, \ldots, a_n = b]$ be a partition of $[a, b]$. From the definition of integrals with Riemann sums, the integral

$$\int_a^b F(t)dt \quad \text{is approximated by the Riemann sum} \quad \sum_{k=0}^{n-1} F(a_k)(a_{k+1} - a_k)$$

whenever $\max(a_{k+1} - a_k)$ is small, and

$$\int_a^b |F(t)|dt \quad \text{is approximated by} \quad \sum_{k=0}^{n-1} |F(a_k)|(a_{k+1} - a_k).$$

The proof is concluded by using the triangle inequality.]

Solution. Given $\epsilon > 0$ there exists $\delta > 0$ such that for any partition of size $< \delta$ we have the following two inequalities:

$$\left| \int_a^b F - \sum_{k=0}^{n-1} F(a_k)(a_{k+1} - a_k) \right| < \epsilon \tag{1}$$

$$\left| \int_a^b |F| - \sum_{k=0}^{n-1} |F(a_k)|(a_{k+1} - a_k) \right| < \epsilon \qquad (2)$$

Then (1) and the triangle inequality imply

$$\left| \int_a^b F \right| \le \left| \sum_{k=0}^{n-1} F(a_k)(a_{k+1} - a_k) \right| + \epsilon \le \sum_{k=0}^{n-1} |F(a_k)|(a_{k+1} - a_k) + \epsilon$$

so that combined with (2) we get

$$\left| \int_a^b F \right| \le \sum_{k=0}^{n-1} |F(a_k)|(a_{k+1} - a_k) + \epsilon \le \int_a^b |F| + 2\epsilon.$$

So for all $\epsilon > 0$ the inequality

$$\left| \int_a^b F \right| \le \int_a^b |F| + 2\epsilon$$

is true hence the inequality $\left| \int_a^b F \right| \le \int_a^b |F|$ holds, as was to be shown.

III.5   The Homotopy Form of Cauchy's Theorem

Exercise III.5.1. *A set S is called* **star-shaped** *if there exists a point z_0 in S such that the line segment between z_0 and any point z in S is contained in S. Prove that a star-shaped set is simply connected, that is, every closed path is homotopic to a point.*

Solution. Consider a closed path $\gamma : [a, b] \to S$. Then by hypothesis we see that we have a homotopy $F_\gamma : [a, b] \times [0, 1] \to S$ defined by

$$F_\gamma(t, u) = (1 - u)\gamma(t) + uz_0.$$

Hence every closed path in S is homotopic to z_0 which implies that S is simply connected.

Exercise III.5.2. *Let U be the open set obtained from* **C** *by deleting the set of real numbers ≥ 0. Prove that U is simply connected.*

Solution. By the previous exercise, it is sufficient to show that U is star-shaped. Let $z_0 = -1$. A picture shows that U is star-shaped, but here is a formal argument. Given $z = x + iy$ consider the segment $[z, z_0]$ parametrized by $\gamma(t) = tz + (1-t)z_0$. The imaginary part of $\gamma(t)$ is given by ty so we see at once that the line segment is contained in U.

Exercise III.5.3. *Let V be the open set obtained from* **C** *by deleting the set of real numbers ≤ 0. Prove that V is simply connected.*

Solution. Argue as in the previous exercise, with $z_0 = 1$.

Exercise III.5.4. *(a) Let U be a simply connected open set and let f be an analytic function on U. Is f(U) simply connected?*
(b) Let H be the upper half-plane, that is, the set of complex numbers $z = x + iy$ such that $y > 0$. Let $f(z) = e^{2\pi iz}$. What is the image $f(H)$? Is $f(H)$ simply connected?

Solution. We first solve (b). If $z = x + iy$, then we see that

$$f(z) = e^{-2\pi y}e^{i(2\pi x)}$$

If $y > 0$, then $0 < e^{-2\pi y} < 1$ so the image of H is the open unit disc minus the origin which is not simply connected. This shows that the image of a simply connected set under an analytic map need not be simply connected. This answers (a) because we see that the image of a simply connected set need not be simply connected.

III.6   Existence of Global Primitives. Definition of the Logarithm

Exercise III.6.1. *Compute the following values when the log is defined by its principal value on the open set U equal to the plane with the positive real axis deleted. (a) $\log i$ (b) $\log(-i)$ (c) $\log(-1 + i)$ (d) i^i (e) $(-i)^i$ (f) $(-1)^i$ (g) $(-1)^{-i}$ (h) $\log(-1 - i)$*

Solution. (a) $\log i = \log e^{i\pi/2} = \log 1 + i\frac{\pi}{2} = i\frac{\pi}{2}$
(b) $\log(-i) = \log e^{3i\pi/2} = 3i\frac{\pi}{2}$
(c) $\log(-1 + i) = \frac{1}{2}\log 2 + i\frac{3\pi}{4}$
(d) $i^i = e^{-\pi/2}$
(e) $(-i)^i = e^{-3\pi/2}$
(f) $(-1)^i = e^{-\pi}$
(g) $(-1)^{-i} = e^{\pi}$
(h) $\log(-1 - i) = \frac{1}{2}\log 2 + i\frac{5\pi}{4}$

Exercise III.6.2. *Compute the values of the same expressions as in Exercise 1 (except (f) and (g)) when the open set consists of the plane from which the negative real axis has been deleted. Then take $-\pi < \theta < \pi$.*

Solution. (a) $\log i = i\frac{\pi}{2}$
(b) $\log(-i) = \log e^{3i\pi/2} = -i\frac{\pi}{2}$
(c) $\log(-1 + i) = \frac{1}{2}\log 2 + i\frac{3\pi}{4}$
(d) $i^i = e^{-\pi/2}$
(e) $(-i)^i = e^{\pi/2}$
(h) $\log(-1 - i) = \frac{1}{2}\log 2 - i\frac{3\pi}{4}$

Exercise III.6.3. *Let U be the plane with the negative real axis deleted. Let $y > 0$. Find the limit*

$$\lim_{y \to 0} [\log(a + iy) - \log(a - iy)]$$

where $a > 0$, and also where $a < 0$.

Solution. First suppose $a > 0$. Let $r_y = (a^2 + y^2)^{1/2}$. Then for small $y > 0$ we can write $a + iy = r_y e^{i\epsilon_y}$ and $a - iy = r_y e^{-i\epsilon_y}$ with $0 < \epsilon_y < \pi/2$. Hence

$$\log(a + iy) - \log(a - iy) = \log r_y + i\epsilon_y - \log r_y - i(-\epsilon_y) = 2i\epsilon_y$$

and since $\epsilon_y \to 0$ as $y \to 0$ we find that

$$\lim_{y \to 0} [\log(a + iy) - \log(a - iy)] = 0.$$

Now suppose $a < 0$. In this case we can write $a + iy = r_y e^{i(\pi - \epsilon_y)}$ and $a - iy = r_y e^{i(-\pi + \epsilon_y)}$ so that arguing as above we get

$$\log(a + iy) - \log(a - iy) = 2\pi i - 2i\epsilon_y,$$

therefore

$$\lim_{y \to 0} [\log(a + iy) - \log(a - iy)] = 2\pi i.$$

Exercise III.6.4. *Let U be the plane with the positive real axis deleted. Find the limit*

$$\lim_{y \to 0} [\log(a + iy) - \log(a - iy)]$$

where $a < 0$, and also where $a > 0$.

Solution. We argue as we did in the previous exercise. When $a > 0$, write $a + iy = r_y e^{i\epsilon_y}$ and $a - iy = r_y e^{i(2\pi - \epsilon_y)}$ so that

$$\lim_{y \to 0} [\log(a + iy) - \log(a - iy)] = -2\pi i.$$

When $a < 0$, we write $a + iy = r_y e^{i(\pi - \epsilon_y)}$ and $a - iy = r_y e^{i(\pi + \epsilon_y)}$ and therefore

$$\lim_{y \to 0} [\log(a + iy) - \log(a - iy)] = 0.$$

Exercise III.6.5. *Over what kind of open sets could you define an analytic function $z^{1/3}$, or more generally $z^{1/n}$ for any positive integer n? Give examples, taking the open set to be as "large" as possible.*

Solution. We can define an analytic function $z^{1/n}$ over any simply connected open set not containing the origin. Indeed, on such a set the log is well defined. Then we define

$$z^{1/n} \stackrel{\text{def}}{=} e^{\frac{1}{n} \log z}$$

Exercise III.6.6. *Let U be a simply connected open set. Let f be analytic on U and assume that f(z) ≠ 0 for all z ∈ U. Show that there exists an analytic function g on U such that $g^2 = f$. Does this last assertion remain true if 2 is replaced by an arbitrary positive integer n?*

Solution. We show that the assertion is true for an arbitrary $n > 0$. It is shown in the text that we can define $\log f(z)$ on U. Define g_n by

$$g_n(z) = f(z)^{1/n} = e^{\frac{1}{n}\log f(z)}.$$

Then g_n is a solution to the problem because

$$g_n(z)\cdots g_n(z) = e^{\frac{1}{n}\log f(z)+\cdots+\frac{1}{n}\log f(z)} = e^{\log f(z)}.$$

Exercise III.6.7. *Let U be the upper half plane, consisting of all complex numbers z = x + iy with y > 0. Let $\varphi(z) = e^{2\pi i z}$. Prove that $\varphi(U)$ is the open unit disc from which the origin has been deleted.*

Solution. If $z = x + iy$ we get

$$\varphi(z) = e^{-2\pi y}e^{2\pi i x}.$$

Since x ranges over **R** and $y > 0$ it is clear that $\varphi(U)$ is the open unit disc minus the origin.

Exercise III.6.8. *Let U be the open set obtained by deleting 0 and the negative real axis from the complex numbers. For an integer $m \geq 1$ define*

$$L_{-m}(z) = \left(\log z - \left(1 + \frac{1}{2} + \cdots + \frac{1}{m}\right)\right)\frac{z^m}{m!}.$$

Show that $L'_{-m}(z) = L_{-m+1}(z)$, and that $L'_{-1}(z) = \log z$. Thus L_{-m} is an m-fold integral of the logarithm.

Solution. Using the rule for differentiating a product we find that

$$L'_{-m}(z) = \frac{1}{z}\frac{z^m}{m!} + \left(\log z - \left(1 + \frac{1}{2} + \cdots + \frac{1}{m}\right)\right)\frac{z^{m-1}}{(m-1)!}$$

$$= \left(\frac{1}{m} + \log z - \left(1 + \frac{1}{2} + \cdots + \frac{1}{m}\right)\right)\frac{z^{m-1}}{(m-1)!} = L_{-m+1}(z).$$

In particular we have $L_{-1}(z) = (\log z - 1)z$ so

$$L'_{-1}(z) = \frac{1}{z}z + (\log z - 1) = \log z$$

hence L_{-m} is an m-fold integral of the logarithm.

III.7 The Local Cauchy Formula

Exercise III.7.1. *Find the integrals over the unit circle γ: (a) $\int_\gamma \frac{\cos z}{z}dz$ (b) $\int_\gamma \frac{\sin z}{z}dz$ (c) $\int_\gamma \frac{\cos(z^2)}{z}dz$.*

Solution. (a) We use the local Cauchy formula on $\overline{D}(0, 1)$. Let $f(z) = \cos z$ and $z_0 = 0$. Then

$$f(z_0) = \frac{1}{2\pi i} \int_\gamma \frac{\cos z}{z} dz.$$

Therefore $\int_\gamma \frac{\cos z}{z} dz = 2\pi i$.

(b) The same argument as in (a) shows that $\int_\gamma \frac{\sin z}{z} dz = 0$.

(c) Arguing like in (a) we find $\int_\gamma \frac{\cos(z^2)}{z} dz = 2\pi i$.

Exercise III.7.2. *Write out completely the proof of Theorem 7.7 to see that all the steps in the proof of Theorem 7.3 apply.*

Solution. Let $z_0 \in U$ and z_0 not on γ. Since the image of γ is compact there is a minimum distance between z_0 and points on γ. Select $0 < R < \text{dist}(z_0, \gamma)$ and take R also small enough that the closed disc $\overline{D}(z_0, R)$ is contained in U. Select $0 < s < R$. We write

$$\frac{1}{\zeta - z} = \frac{1}{\zeta - z_0} \left(\frac{1}{1 - \frac{z - z_0}{\zeta - z_0}} \right)$$

$$= \frac{1}{\zeta - z_0} \left(1 + \frac{z - z_0}{\zeta - z_0} + \left(\frac{z - z_0}{\zeta - z_0} \right)^2 + \cdots \right).$$

This series converges absolutely and uniformly for $|z - z_0| \leq s$ because

$$\left| \frac{z - z_0}{\zeta - z_0} \right| \leq \frac{s}{R} < 1.$$

The function g is continuous on γ so it is bounded. By Theorem 2.4 of Chapter III, we can integrate term by term and we find

$$f(z) = \sum_{n=0}^\infty a_n (z - z_0)^n$$

where

$$a_n = \int_\gamma \frac{g(\zeta)}{(\zeta - z_0)^{n+1}} d\zeta.$$

This proves that f is analytic, and that

$$f^{(n)}(z_0) = n! \int_\gamma \frac{g(\zeta)}{(\zeta - z_0)^{n+1}} d\zeta$$

thereby concluding the proof of Theorem 7.7.

Exercise III.7.3. *Prove Corollary 7.4, that is, prove:*

Corollary. *Let f be an entire function, and let $\|f\|_R$ be its sup norm on the circle of radius R. Suppose that there exists a constant C and a positive integer k such that*

$$\|f\|_R \leq C R^k$$

for arbitrarily large R. Then f is a polynomial of degree $\leq k$.

Solution. Let $n > k$. The estimate $|a_n| \leq \|f\|_R / R^n$ of the coefficients in the power series expansion of f gives in our particular case the inequality

$$|a_n| \leq C R^k / R^n$$

which holds for all large R. Letting $R \to \infty$ we get that $a_n = 0$ for all $n > k$. So in the power series expansion of f, all terms of degree $> k$ are equal to 0, whence f is a polynomial of degree $\leq k$.

IV

Winding Numbers and Cauchy's Theorem

IV.2   The Global Cauchy Theorem

Exercise IV.2.1. (a) *Show that the association* $f \mapsto f'/f$ *(where f is holomorphic) sends products to sums.*
(b) *If* $P(z) = (z - a_1) \cdots (z - a_n)$, *where* $a_1, \ldots, a_n$ *are the roots, what is* P'/P?
(c) *Let* γ *be a closed path such that none of the roots of P lie on γ. Show that*

$$\frac{1}{2\pi i} \int_\gamma (P'/P)(z)dz = W(\gamma, a_1) + \cdots + W(\gamma, a_n).$$

Solution. (a) The product rule for differentiation implies that

$$\frac{(fg)'}{fg} = \frac{f'g}{fg} + \frac{fg'}{fg} = \frac{f'}{f} + \frac{g'}{g},$$

so the association $f \mapsto f'/f$ sends products to sums.
(b) Let us write $\Phi(f) = f'/f$. Part (a) implies that $\Phi\left(\prod f_k\right) = \sum \Phi(f_k)$ so if $f_k = z - a_k$, then $(f_k'/f_k)(z) = 1/(z - a_k)$ and therefore

$$\Phi(P)(z) = \frac{P'}{P}(z) = \sum_{k=1}^{n} \frac{1}{z - a_k}.$$

(c) By definition

$$W(\gamma, a_k) = \frac{1}{2\pi i} \int_\gamma \frac{dz}{z - a_k},$$

so using part (a) we find

$$\frac{1}{2\pi i}\int_\gamma (P'/P)(z)dz = \frac{1}{2\pi i}\sum_{k=1}^n \int_\gamma \frac{dz}{z-a_k} = \sum_{k=1}^n W(\gamma, a_k)$$

as was to be shown.

Exercise IV.2.2. *Let $f(z) = (z - z_0)^m h(z)$, where h is analytic on an open set U, and $h(z) \neq 0$ for all $z \in U$. Let γ be a closed path homologous to 0 in U, and such that z_0 does not lie on γ. Prove that*

$$\frac{1}{2\pi i}\int_\gamma \frac{f'(z)}{f(z)}dz = W(\gamma, z_0)m.$$

Solution. We have

$$\frac{f'(z)}{f(z)} = \frac{m}{z-z_0} + \frac{h'(z)}{h(z)}.$$

The hypothesis that h does not vanish on U implies that h'/h is holomorphic on U and therefore by Cauchy's theorem its integral along γ is 0. So

$$\frac{1}{2\pi i}\int_\gamma \frac{f'(z)}{f(z)}dz = \frac{1}{2\pi i}\int_\gamma \frac{m}{z-z_0}dz = W(\gamma, z_0)m.$$

Exercise IV.2.3. *Let U be a simply connected open set and let $z_1, \ldots, z_n$ be points of U. Let $U^* = U - \{z_1, \ldots, z_n\}$ be the set obtained from U by deleting the points $z_1, \ldots, z_n$. Let f be analytic on U^*. Let γ_k be a small circle centered at z_k and let*

$$a_k = \frac{1}{2\pi i}\int_{\gamma_k} f(\zeta)d\zeta.$$

Let $h(z) = f(z) - \sum a_k/(z - z_k)$. Prove that there exists an analytic function H on U^* such that $H' = h$.

Solution. Fix a point $w \in U^*$. Given a path γ in U^* from w to a point $z \in U^*$ define

$$H_\gamma(z) = \int_\gamma h(\zeta)d\zeta.$$

We claim that this function is independent of the path chosen from w to z. Indeed, suppose η is another path from w to z. Then we have

$$H_\gamma(z) - H_\eta(z) = \int_\mu h(\zeta)d\zeta,$$

where μ is a closed curve in U^*. Since U is simply connected, the path μ is homologous to 0 in U. By Cauchy's theorem, we see that if m_k denotes the winding number of μ with respect to z_k, then

$$\int_\mu h(\zeta)d\zeta = \sum m_k \int_{\gamma_k} h(\zeta)d\zeta.$$

By construction, we have

$$\int_\gamma h(\zeta)d\zeta = 0$$

which proves our claim. So we may use the notation H to denote the function defined above. The standard argument then shows that H is analytic and that $H' = h$. This concludes the proof.

V
Applications of Cauchy's Integral Formula

V.1 Uniform Limits of Analytic Functions

Exercise V.1.1. *Let f be analytic on an open set U, let $z_0 \in U$ and $f'(z_0) \neq 0$. Show that*

$$\frac{2\pi i}{f'(z_0)} = \int_C \frac{1}{f(z) - f(z_0)} dz,$$

where C is a small circle centered at z_0.

Solution. For z near z_0 we can write $f(z) - f(z_0) = a_1(z - z_0) + a_2(z - z_0)^2 + \cdots$ with $a_1 = f'(z_0) \neq 0$. So

$$f(z) - f(z_0) = a_1(z - z_0)\left(1 + \frac{a_2}{a_1}(z - z_0) + \cdots\right)$$

which after inverting the expression in parenthesis implies that on a small disc around z_0 the function given by $\frac{z - z_0}{f(z) - f(z_0)}$ is analytic, thus by Cauchy's formula we get

$$\frac{1}{2\pi i} \int_C \frac{dz}{f(z) - f(z_0)} = \frac{1}{a_1} = \frac{1}{f'(z_0)}.$$

Exercise V.1.2. *Weierstrass' theorem for a real interval $[a, b]$ states that a continuous function can be uniformly approximated by polynomials. Is this conclusion still true for the closed unit disc, i.e., can every continuous function on the disc be uniformly approximated by polynomials?*

Solution. Since polynomials are holomorphic, the uniform limit of a sequence of polynomials is holomorphic. However, not every continuous function on the disc is holomorphic ($z \mapsto \bar{z}$ for example), so the conclusion of Weierstrass' theorem is false for the closed unit disc.

Exercise V.1.3. *Let $a > 0$. Show that each of the following series represents a holomorphic function:*
(a) $\sum_{n=1}^{\infty} e^{-an^2z}$ for $\mathrm{Re}(z) > 0$;
(b) $\sum_{n=1}^{\infty} \frac{e^{-anz}}{(a+n)^2}$ for $\mathrm{Re}(z) > 0$;
(c) $\sum_{n=1}^{\infty} \frac{1}{(a+n)^z}$ for $\mathrm{Re}(z) > 1$.

Solution. (a) Let $c > 0$, $f_n(z) = e^{-an^2z}$ and $z = x + iy$ where x and y are real numbers. Then

$$|e^{-an^2z}| = e^{-an^2x}.$$

If $x = \mathrm{Re}(z) \geq c$, then $e^{-an^2x} \leq e^{-an^2c} \leq e^{-anc}$ and the geometric series $\sum (e^{-ac})^n$ converges for $c > 0$ so we conclude that the series $\sum e^{-an^2z}$ converges uniformly for $\mathrm{Re}(z) \geq c$. Clearly, the functions f_n are holomorphic so the series $\sum e^{-an^2z}$ defines a holomorphic function on $\mathrm{Re}(z) > 0$.

(b) Let $f_n(z) = e^{-anz}/(a+n)^2$. The functions f_n are holomorphic and for $\mathrm{Re}(z) > 0$ we have $|e^{-anz}| < 1$ which implies that $|f_n(z)| \leq 1/n^2$. The convergence of $\sum 1/n^2$ implies that the series $\sum e^{-anz}/(a+n)^2$ defines a holomorphic function for $\mathrm{Re}(z) > 0$.

(c) Let $f_n(z) = 1/(a+n)^z$. Then $|f_n(z)| = 1/(a+n)^x$, so if $\mathrm{Re}(z) \geq 1 + \epsilon$, for $\epsilon > 0$, we get that $|f_n(z)| \leq 1/n^{1+\epsilon}$. Since $\sum 1/n^{1+\epsilon}$ converges, we conclude that the function $\sum 1/(a+n)^z$ is holomorphic for $\mathrm{Re}(z) > 1$.

Exercise V.1.4. *Show that each of the two series converges uniformly on each closed disc $|z| \leq c$ with $0 < c < 1$:*

$$\sum_{n=1}^{\infty} \frac{nz^n}{1-z^n} \quad and \quad \sum_{n=1}^{\infty} \frac{z^n}{(1-z^n)^2}.$$

Solution. Let $a_n(z) = nz^n/(1-z^n)$. We have $|1-z^n| \geq 1 - |z^n| \geq 1 - c^n$ so that

$$|a_n(z)| \leq n\frac{c^n}{1-c^n}.$$

For all sufficiently large n, $1 - c^n \geq 1/2$, so for all large n we have $|a_n(z)| \leq 2nc^n$. The ratio test implies at once that $\sum nc^n$ converges, so the series $\sum nz^n/(1-z^n)$ converges uniformly on each closed disc $|z| \leq c$ with $0 < c < 1$.

Consider the series $\sum z^n/(1-z^n)^2$, and let $a_n(z) = z^n/(1-z^n)^2$. Estimating $|a_n(z)|$ as we did for the first series we find that for large n the inequality

$$|a_n(z)| \leq 4c^n$$

holds. But $\sum c^n$ converges, so $\sum z^n/(1-z^n)^2$ converges uniformly on each closed disc $|z| \leq c$ with $0 < c < 1$.

Exercise V.1.5. *Prove that the two series in Exercise 4 are actually equal. [Hint: Write each one in a double series and reverse the order of summation.]*

Solution. We use the fact that $\frac{1}{1-z^n} = \sum_{k=0}^{\infty}(z^n)^k$ to get

$$\sum_{n=1}^{\infty}\frac{nz^n}{1-z^n} = \sum_{n=1}^{\infty}nz^n\sum_{k=0}^{\infty}(z^n)^k = \sum_{n=1}^{\infty}n\sum_{k=0}^{\infty}(z^n)^{k+1}$$

$$= \sum_{n=1}^{\infty}\sum_{j=1}^{\infty}n(z^n)^j = \sum_{j=1}^{\infty}\sum_{n=1}^{\infty}n(z^j)^n.$$

For the second series we get

$$\sum_{n=1}^{\infty}\frac{z^n}{(1-z^n)^2} = \sum_{n=1}^{\infty}z^n\left[\sum_{k=0}^{\infty}(z^n)^k \cdot \sum_{k=0}^{\infty}(z^n)^k\right] = \sum_{n=1}^{\infty}z^n\sum_{k=0}^{\infty}(k+1)(z^n)^k$$

$$= \sum_{n=1}^{\infty}\sum_{k=0}^{\infty}(k+1)(z^n)^{k+1} = \sum_{n=1}^{\infty}\sum_{j=1}^{\infty}j(z^n)^j.$$

It is now clear that both series are equal.

Exercise V.1.6 (Dirichlet Series). *Let $\{a_n\}$ be a sequence of complex numbers. Show that the series $\sum a_n/n^s$, if it converges at all for some complex s, converges absolutely in a right half-plane $\mathrm{Re}(s) > \sigma_0$, and uniformly in $\mathrm{Re}(s) > \sigma_0 + \epsilon$ for every $\epsilon > 0$. Show that the series defines an analytic function in this half plane. The number σ_0 is called the **abscissa of convergence**.*

Solution. Suppose that the series converges for some s_0. Let $\sigma_0 = \mathrm{Re}(s_0)+1$ and suppose that $\mathrm{Re}(s) > \sigma_0 + \epsilon$. The series $\sum a_n/n^{s_0}$ converges, so for all large n we have the inequality

$$|a_n/n^{s_0}| = |a_n|/n^{\mathrm{Re}(s_0)} \leq 1$$

which implies that for all large n we have

$$|a_n/n^s| \leq |a_n|/n^{\sigma_0+\epsilon} = |a_n|/n^{\mathrm{Re}(s_0)+1+\epsilon} \leq 1/n^{1+\epsilon}.$$

Conclude.

Exercise V.1.7. *Let f be analytic on a closed disc $\overline{D}$ of radius $b > 0$, centered at z_0. Show that*

$$\frac{1}{\pi b^2}\int\int_D f(x+iy)dydx = f(z_0).$$

[Hint: Use polar coordinates and Cauchy's formula. Without loss of generality, you may assume that $z_0 = 0$. Why?]

Solution. Let $g(z) = f(z+z_0)$. A linear change of variables shows that

$$\int\int_D f(x+iy)dxdy = \int\int_{D(0,b)} g(x+iy)dxdy,$$

and $g(0) = f(z_0)$, so we may assume that $z_0 = 0$. If $0 < r < b$ and C_r denotes the circle centered at the origin of radius r, Cauchy's formula implies.

$$f(0) = \frac{1}{2\pi i} \int_{C_r} \frac{f(\zeta)}{\zeta} d\zeta.$$

We parametrize C_r by $re^{i\theta}$ with $\theta \in [0, 2\pi]$, so that

$$f(0) = \frac{1}{2\pi i} \int_0^{2\pi} \frac{f(re^{i\theta})}{re^{i\theta}} ire^{i\theta} d\theta = \frac{1}{2\pi} \int_0^{2\pi} f(re^{i\theta}) d\theta.$$

We can now multiply both sides of the above equality by r and integrate with respect to r from 0 to b. Interchanging the order of integration we obtain

$$f(0) \int_0^b r\,dr = \frac{1}{2\pi} \int_0^{2\pi} \int_0^b f(re^{i\theta}) r\,dr\,d\theta.$$

Evaluating the integral on the left and changing variables in the integral on the right (from polar to rectangular), we get after some simplifications

$$f(0) = \frac{1}{\pi b^2} \int\int_{D(0,b)} f(x+iy)dy\,dx,$$

which is the desired formula.

Exercise V.1.8. *Let D be the unit disc and let S be the unit square, that is, the set of complex numbers z such that $0 < \text{Re}(z) < 1$ and $0 < \text{Im}(z) < 1$. Let $f : D \to S$ be an analytic isomorphism such that $f(0) = (1+i)/2$. Let u, v be the real and imaginary parts of f respectively. Compute the integral*

$$\int\int_D \left[\left(\frac{\partial u}{\partial x}\right)^2 + \left(\frac{\partial u}{\partial y}\right)^2\right]dx\,dy.$$

Solution. By the Cauchy–Riemann equations we find that

$$\left(\frac{\partial u}{\partial x}\right)^2 + \left(\frac{\partial u}{\partial y}\right)^2 = \Delta_f,$$

where Δ_f is the Jacobian determinant of f. Applying the change of variable formula we get

$$\int\int_S dx\,dy = \int\int_D |f'(z)|^2 dx\,dy$$

so the desired integral is equal to the area of the unit square namely

$$\int\int_D \left[\left(\frac{\partial u}{\partial x}\right)^2 + \left(\frac{\partial u}{\partial y}\right)^2\right]dx\,dy = 1.$$

Exercise V.1.9. *(a) Let f be an analytic isomorphism on the unit disc D, and let*

$$f(z) = \sum_{n=1}^{\infty} a_n z^n$$

be its power series expansion. Prove that

$$\text{area } f(D) = \pi \sum_{n=1}^{\infty} n|a_n|^2.$$

(b) Suppose that f is an analytic isomorphism on the closed unit disc $\overline{D}$, and that $|f(z)| \geq 1$ if $|z| = 1$, and $f(0) = 0$. Prove that area $f(D) \geq \pi$.

Solution. (a) Applying the change of variable formula together with the Cauchy–Riemann equations we find

$$\text{area } f(D) = \int\int_{f(D)} dx\,dy = \int\int_{D} |f'(z)|^2 dx\,dy.$$

Switching to polar coordinates we get

$$\text{area } f(D) = \int_0^1 \int_0^{2\pi} r|f'(re^{i\theta})|^2 d\theta\,dr.$$

However, $f'(z) = \sum_{n=1}^{\infty} na_n z^{n-1}$ thus

$$|f'(re^{i\theta})|^2 = (a_1 + 2a_2 re^{i\theta} + \cdots)(\overline{a}_1 + 2\overline{a}_2 re^{-i\theta} + \cdots).$$

For all nonzero integers n a direct calculation shows that $\int_0^{2\pi} e^{ni\theta} d\theta = 0$ hence

$$\text{area } f(D) = \int_0^1 \int_0^{2\pi} r \left(\sum_{n=1}^{\infty} n^2 |a_n|^2 r^{2n-2} \right) d\theta\,dr$$

$$= 2\pi \sum_{n=1}^{\infty} \left(n^2 |a_n|^2 \int_0^1 r^{2n-1} dr \right).$$

The desired formula drops out after evaluation of the last integral.

(b) We are given that $|f(e^{i\theta})| \geq 1$ for all real θ. So

$$\int_0^{2\pi} |f(e^{i\theta})|^2 d\theta \geq 2\pi.$$

Since $f(0) = 0$, the constant term of the power series of f at 0 is 0. Using the fact that $\int_0^{2\pi} e^{ni\theta} d\theta = 0$ for all nonzero integers n, we find that

$$\int_0^{2\pi} |f(e^{i\theta})|^2 d\theta = \int_0^{2\pi} \sum_{n=1}^{\infty} |a_n|^2 d\theta = 2\pi \sum_{n=1}^{\infty} |a_n|^2.$$

Therefore combined with the previous inequality we get $\sum_{n=1}^{\infty} |a_n|^2 \geq 1$. The result now follows from part (a)

$$\text{area } f(D) = \pi \sum_{n=1}^{\infty} n|a_n|^2 \geq \pi \sum_{n=1}^{\infty} |a_n|^2 \geq \pi.$$

Exercise V.1.10. *Let f be analytic on the unit disc D and assume that $\int\int_D |f|^2 dx dy$ exists. Let*

$$f(z) = \sum_{n=0}^{\infty} a_n z^n.$$

Prove that

$$\frac{1}{2\pi} \int\int_D |f(z)|^2 dx dy = \sum_{n=0}^{\infty} |a_n|^2/(2n+2).$$

Solution. To compute the integral we use polar coordinates,

$$\int\int_D |f(z)|^2 dx dy = \int_0^{2\pi} \int_0^1 |f(re^{i\theta})|^2 r dr d\theta = \int_0^1 \int_0^{2\pi} |f(re^{i\theta})|^2 r d\theta dr.$$

Arguing like in the preceding exercise we find that

$$\int_0^{2\pi} |f(re^{i\theta})|^2 d\theta = 2\pi \sum_{n=0}^{\infty} |a_n|^2 r^{2n}$$

so that

$$\int_0^1 \int_0^{2\pi} |f(re^{i\theta})|^2 r d\theta dr = 2\pi \sum_{n=0}^{\infty} |a_n|^2/(2n+2)$$

as was to be shown.

For the next exercise, recall that a norm $\| \cdot \|$ on a space of functions associates to each function f a real number ≥ 0, satisfying the following conditions:
N 1. *We have $\|f\| = 0$ if and only if $f = 0$.*
N 2. *If c is a complex number, then $\|cf\| = |c| \|f\|$.*
N 3. *$\|f + g\| \leq \|f\| + \|g\|$.*

Exercise V.1.11. *Let A be the closure of a bounded open set in the plane. Let f, g be continuous functions on A. Define their scalar product*

$$\langle f, g \rangle = \int\int_A f(z)\overline{g(z)} dy dx$$

and define the associated L^2-norm by its square,

$$\|f\|_2^2 = \int\int_A |f(z)|^2 dy dx.$$

*Show that $\|f\|_2$ does define a norm. Prove the **Schwarz inequality***

$$|\langle f, g \rangle| \leq \|f\|_2 \|g\|_2.$$

On the other hand, define

$$\|f\|_1 = \int\int_A |f(z)| dy dx.$$

Show that $f \mapsto \|f\|_1$ is a norm on the space of continuous functions on A, called the L^1-norm. This is just preliminary. Prove:
(a) Let $0 < s < R$. Prove that there exist constants C_1, C_2 having the following property. If f is analytic on a closed disc $\overline{D}$ of radius R, then

$$\|f\|_s \leq C_1\|f\|_{1,R} \leq C_2\|f\|_{2,R},$$

where $\| \cdot \|_s$ is the sup norm on the closed disc of radius s, and the L^1, L^2 norms refer to the integral over the disc of radius R.
(b) Let $\{f_n\}$ be a sequence of holomorphic functions on an open set U, and assume that this sequence is L^2-Cauchy. Show that it converges uniformly on compact subsets of U.

Solution. First we prove the preliminaries. The Schwarz inequality is a standard result in linear algebra of hermitian scalar products. Let $\alpha = \langle g, g \rangle$ and $\beta = -\langle f, g \rangle$. Then

$$0 \leq \langle \alpha f + \beta g, \alpha f + \beta g \rangle$$
$$= \alpha\overline{\alpha}\langle f, f \rangle + \beta\overline{\alpha}\langle g, f \rangle + \alpha\overline{\beta}\langle f, g \rangle + \beta\overline{\beta}\langle g, g \rangle.$$

Note that $\alpha = \|g\|_2^2$. Substituting the values for α, β we obtain

$$0 \leq \|g\|_2^4\|f\|_2^2 - 2\|g\|_2^2\langle f, g\rangle\overline{\langle f, g\rangle} + \|g\|_2^2\langle f, g\rangle\overline{\langle f, g\rangle}.$$

But $\langle f, g\rangle\overline{\langle f, g\rangle} = |\langle f, g\rangle|^2$ hence

$$\|g\|_2^2|\langle f, g\rangle|^2 \leq \|g\|_2^4\|f\|_2^2.$$

Conclude the proof by considering both cases $\|g\|_2 = 0$ and $\|g\|_2 \neq 0$.

We now show that $\| \cdot \|_2$ is a norm. Clearly, $\|f\|_2 \geq 0$ for all f. Suppose that f is not identically zero. Then by continuity, $|f(z)| > \delta > 0$ for all z in some ball, so $\|f\|_2 > 0$. The second condition is obvious, Finally for the triangle inequality we use the Schwarz inequality. We have

$$\|f + g\|_2^2 = \langle f + g, f + g \rangle = \langle f, f \rangle + \langle f, g \rangle + \langle g, f \rangle + \langle g, g \rangle$$

but $\langle f, g \rangle + \langle g, f \rangle = 2\,\mathrm{Re}\langle f, g\rangle \leq 2|\langle f, g\rangle|$, so

$$\|f + g\|_2^2 \leq \|f\|_2^2 + 2|\langle f, g\rangle| + \|g\|_2^2$$
$$\leq \|f\|_2^2 + 2\|f\|_2\|g\|_2 + \|g\|_2^2$$
$$\leq (\|f\|_2 + \|g\|_2)^2.$$

The triangle inequality follows from taking square roots on both sides.

Finally, $\| \cdot \|_1$ is a norm. The first two conditions are obvious, and the triangle inequality follows from the triangle inequality for the standard absolute value of complex numbers.

(a) These two estimates show that the integral norms dominate the sup norm. To prove the first inequality, we proceed as follows. Fix a real number r_0 such that $s < r_0 < R$. There exists a constant $\delta > 0$ such that $|z - \zeta| > \delta$ for all z in the closed disc of radius s and all ζ on a circle of radius r with $r_0 < r < R$. In fact, we may take $\delta = r_0 - s$. Then, applying Cauchy's formula we obtain

$$f(z) = \frac{1}{2\pi i} \int_{C_r} \frac{f(\zeta)}{\zeta - z} d\zeta.$$

Writing $\zeta = re^{i\theta}$, putting absolute values and using the above observation we get

$$|f(z)| \leq \frac{1}{2\pi\delta} \int_0^{2\pi} |f(re^{i\theta})| r \, d\theta.$$

Integrating both sides with respect to r from r_0 to R, and using polar coordinates we find

$$(R - r_0)|f(z)| \leq C \int_{r_0 < |z| < R} |f| \leq C \int_{\overline{D}(0,R)} |f|$$

which completes the proof. The second inequality follows from applying the Schwarz inequality taking one function to be $|f|$ and the other to be 1.

(b) Let K be a compact subset of U. For each point z in K, choose $r_z > 0$ such that $D(z, 2r_z) \subset U$. Then $\bigcup_{z \in K} D(z, r_z)$ is an open cover of K from which we can select a finite subcover $\bigcup_{i=1}^p D(z_i, r_i)$. By part (a) we have $\|f\|_{r_i} \leq C_i \|f\|_{2,2r_i}$ for some constant C_i. Clearly $\|f\|_{2,2r_i} \leq \|f\|_{2,\overline{U}}$ so $\|f\|_{r_i} \leq C_i \|f\|_{2,\overline{U}}$. If $\{f_n\}$ is L^2-Cauchy, then given $\epsilon > 0$ there exists N such that $\|f_n - f_m\|_{2,\overline{U}} < \epsilon$ for all $n, m > N$. So $\|f_n - f_m\|_{r_i} < C_i\epsilon$ for all $i = 1, \ldots, p$ and therefore $\|f_n - f_m\|_K < C\epsilon$ for some positive constant C (actually, $C = \sum C_i$ will do). This proves that $\{f_n\}$ is uniformly Cauchy on K and therefore $\{f_n\}$ converges uniformly on compact subsets of U.

Exercise V.1.12. *Let U, V be open discs centered at the origin. Let $f = f(z, w)$ be a continuous function on the product $U \times V$, such that for each w the function $z \mapsto f(z, w)$ and for each z the function $w \mapsto f(z, w)$ are analytic on U and V, respectively. Show that f has a power series expansion*

$$f(z, w) = \sum a_{mn} z^m w^n$$

which converges absolutely and uniformly for $|z| \leq r$ and $|w| \leq r$, for some positive number r. [Hint: Apply Cauchy's formula for derivatives twice, with respect to the two variables to get an estimate of the coefficients a_{mn}.] Generalize to several variables instead of two variables.

Solution. Select $R > 0$ such that $\overline{D}(0, R) \subset U$ and $\overline{D}(0, R) \subset V$. Fixing w, Cauchy's formula implies

$$f(z, w) = \frac{1}{2\pi i} \int_{C_R} \frac{f(\zeta, w)}{\zeta - z} d\zeta$$

for $z \in \overline{D}(0, R)$. For fixed ζ, we have

$$f(\zeta, w) = \frac{1}{2\pi i} \int_{C_R} \frac{f(\zeta, \xi)}{\xi - w} d\xi$$

for $w \in \overline{D}(0, R)$. Hence

$$f(z, w) = \left(\frac{1}{2\pi i}\right)^2 \int_{C_R} \int_{C_R} \frac{f(\zeta, \xi)}{(\zeta - z)(\xi - w)} d\xi d\zeta.$$

Since f is continuous, we can apply Fubini's theorem to transform the iterated integral into an integral over $C_R \times C_R$. We get

$$f(z, w) = \left(\frac{1}{2\pi i}\right)^2 \int_{C_R \times C_R} \frac{f(\zeta, \xi)}{(\zeta - z)(\xi - w)} d\xi d\zeta.$$

Now suppose that $0 < r < R$ and that $|z| \leq r$ and $|w| \leq r$. Then we can write

$$\frac{1}{(\zeta - z)(\xi - w)} = \frac{1}{\zeta(1 - z/\zeta)} \frac{1}{\xi(1 - w/\xi)}$$

$$= \frac{1}{\zeta} \left(\sum_{m=0}^{\infty} \frac{z^m}{\zeta^m}\right) \frac{1}{\xi} \left(\sum_{n=0}^{\infty} \frac{w^n}{\xi^n}\right)$$

$$= \sum_{m,n=0}^{\infty} \frac{z^m w^n}{\zeta^{m+1} \xi^{n+1}}.$$

This series is absolutely and uniformly convergent whenever $|z| \leq r$ and $|w| \leq r$. Since f is bounded on $C_R \times C_R$ we can integrate term by term and we get

$$f(z, w) = \sum_{m,n=0}^{\infty} a_{mn} z^m w^n$$

where

$$a_{mn} = \left(\frac{1}{2\pi i}\right)^2 \int_{C_R \times C_R} \frac{f(\zeta, \xi)}{\xi^{n+1} \zeta^{m+1}} d\xi d\zeta.$$

Note that we have the estimates

$$|a_{mn}| \leq \frac{1}{4\pi^2} \frac{\|f\|_{C_R \times C_R}}{R^{n+m+2}} 4\pi^2 R^2 = \frac{\|f\|_{C_R \times C_R}}{R^{n+m}}.$$

where $\| \cdot \|$ denotes the sup norm.

The same argument shows that for n variables, we have the expansion

$$f(z_1, \ldots, z_n) = \sum_{i_1} \cdots \sum_{i_n} a_{i_1 \cdots i_n} z_1^{i_1} \cdots z_n^{i_n}$$

which converges for $|z_i| \leq r < R$. Moreover the formula that gives the coefficients is

$$a_{i_1 \cdots i_n} = \int_{C_R} \cdots \int_{C_R} \frac{f(\zeta_1, \ldots, \zeta_n)}{\zeta_1^{i_1+1} \cdots \zeta_n^{i_n+1}} d\zeta_n \cdots d\zeta_1.$$

We also have the estimate

$$|a_{i_1 \cdots i_n}| \leq \frac{\|f\|_{C_R \times \cdots \times C_R}}{R^{i_1 + \cdots + i_n}}.$$

V.2   Laurent Series

Exercise V.2.1. *Prove that the Laurent series can be differentiated term by term in the usual manner to give the derivative of f on the annulus.*

Solution. We use the notation of the section and of Theorem 2.1. Write $f = f^+ + f^-$. Look at $f^+(z) = \sum_{k=0}^{\infty} a_k z^k$ and let $f_n^+(z) = \sum_{k=0}^{n} a_k z^k$. Then $f_n^+ \to f^+$ uniformly on a slightly smaller annulus, namely on the annulus $s \leq |z| \leq S$. So $(f_n^+)' \to (f^+)'$ uniformly on $s \leq |z| \leq S$ and therefore we can differentiate term by term. The same argument shows that we can differentiate term by term f^- to obtain $(f^-)'$.

Exercise V.2.2. *Let f be holomorphic on the annulus A, defined by*

$$0 < r \leq |z| \leq R.$$

Prove that there exist functions f_1, f_2 such that f_1 is holomorphic for $|z| \leq R$, f_2 is holomorphic for $|z| \geq r$ and

$$f = f_1 + f_2$$

on the annulus.

Solution. Write $f(z) = \sum_{n=-\infty}^{\infty} a_n z^n$. Let $f_1(z) = \sum_{n=0}^{\infty} a_n z^n$ and $f_2(z) = \sum_{n=1}^{\infty} a_{-n} z^{-n}$. We show that f_1 is holomorphic for $|z| \leq R$. Since it is holomorphic in the annulus we must only consider the case when $|z| \leq r$. But then, if $r < S < R$, we have

$$|a_n z^n| \leq |a_n| r^n \leq |a_n| S^n$$

and by assumption the series $\sum_{n \geq 0} |a_n| S^n$ converges so this proves that f_1 is holomorphic for $|z| \leq R$. The same argument shows that f_2 is holomorphic for $|z| \geq r$.

Exercise V.2.3. *Is there a polynomial $P(z)$ such that $P(z)e^{1/z}$ is an entire function? Justify your answer. What is the Laurent expansion of $e^{1/z}$ for $|z| \neq 0$?*

Solution. We know that $e^z = \sum_{n=0}^{\infty} z^n/n!$, so

$$e^{1/z} = \sum_{n=0}^{\infty} \frac{1}{n!} \left(\frac{1}{z}\right)^n.$$

If we have a polynomial $P(z) = a_d z^d + \cdots + a_0$, then the Laurent expansion of $P(z)e^{1/z}$ near zero will have terms of the form $\frac{c_N}{z^N}$ where $c_N \neq 0$ and $N > 0$. To see this, note that the coefficient of $1/z^N$ in the expansion of $P(z)e^{1/z}$ is

$$c_N = \frac{a_0}{N!} + \frac{a_1}{(N+1)!} + \cdots + \frac{a_d}{(N+d)!}.$$

If r is the smallest nonnegative integer such that $a_r \neq 0$, then we see that we can rewrite

$$c_N = \frac{1}{(N+r)!} \left[a_r + \frac{a_{r+1}}{N+r+1} + \cdots + \frac{a_d}{(N+r+1)\cdots(N+d)} \right]$$

and therefore this coefficient is nonzero for all large N. This proves that there is no polynomial $P(z)$ such that $P(z)e^{1/z}$ is entire.

Exercise V.2.4. *Expand the function*

$$f(z) = \frac{z}{1+z^3}$$

(a) in a series of positive powers of z, and
(b) in a series of negative powers of z.
In each case, specify the region in which the expansion is valid.

Solution. (a) For $|z| < 1$, we have

$$f(z) = z\frac{1}{1-(-z^3)} = z\left(1 + (-z^3) + (-z^3)^2 + \cdots\right)$$

so that $f(z) = \sum_{n=0}^{\infty}(-1)^n z^{3n+1}$.
(b) For $|z| > 1$, we write

$$f(z) = \frac{1}{z^2}\frac{1}{1+1/z^3} = \frac{1}{z^2}\left(1 + \left(\frac{-1}{z^3}\right) + \left(\frac{-1}{z^3}\right)^2 + \cdots\right)$$

whence $f(z) = \sum_{n=0}^{\infty}(-1)^n(1/z)^{3n+2}$.

Exercise V.2.5. *Give the Laurent expansions for the following functions:*
(a) $z/(z+2)$ for $|z| > 2$
(b) $\sin 1/z$ for $z \neq 0$
(c) $\cos 1/z$ for $z \neq 0$
(d) $\frac{1}{(z-3)}$ for $|z| > 3$

Solution. (a) $\frac{z}{z+2} = \frac{1}{1+2/z} = \sum_{n=0}^{\infty}(2/z)^n$.
(b) $\sin 1/z = \sum_{n=0}^{\infty}\frac{(-1)^n}{(2n+1)!}\frac{1}{z^{2n+1}}$.
(c) $\cos 1/z = \sum_{n=0}^{\infty}\frac{(-1)^n}{(2n)!}\frac{1}{z^{2n}}$.
(d) $\frac{1}{z-3} = \frac{1}{z}\frac{1}{1-3/z} = \sum_{n=0}^{\infty}\frac{(-3)^n}{z^{n+1}}$.

Exercise V.2.6. *Prove the following expansions:*
(a) $e^z = e + e\sum_{n=1}^{\infty}\frac{1}{n!}(z-1)^n$
(b) $1/z = \sum_{n=0}^{\infty}(-1)^n(z-1)^n$ for $|z-1| < 1$
(c) $1/z^2 = 1 + \sum_{n=1}^{\infty}(n+1)(z+1)^n$ for $|z+1| < 1$

Solution. (a) By definition,

$$e^{z-1} = \sum_{n=0}^{\infty}\frac{1}{n!}(z-1)^n = 1 + \sum_{n=1}^{\infty}\frac{1}{n!}(z-1)^n.$$

However $e^{z-1} = e^z e^{-1}$.
(b) For $|z-1| < 1$ we can use the formula for the geometric series,

$$\frac{1}{z} = \frac{1}{1+(z-1)} = \sum_{n=0}^{\infty}(-1)^n(z-1)^n.$$

(c) The derivative of $-1/z$ is $1/z^2$. For $|z + 1| < 1$ we have

$$\frac{-1}{z} = \frac{1}{1 - (z + 1)},$$

so $-1/z = \sum_{n=0}^{\infty}(z + 1)^n = 1 + \sum_{n=1}^{\infty}(z + 1)^n$ and therefore

$$\frac{1}{z^2} = \sum_{n=1}^{\infty} n(z + 1)^{n-1} = 1 + \sum_{n=1}^{\infty}(n + 1)(z + 1)^n$$

which proves the formula.

Exercise V.2.7. *Expand (a) $\cos z$, (b) $\sin z$ in a power series about $\pi/2$.*

Solution. (a) Just like in the real case we have $\cos z = -\sin(z - \pi/2)$ for all $z \in \mathbf{C}$, so

$$\cos z = \sum_{n=0}^{\infty}(-1)^{n+1}\frac{(z - \pi/2)^{2n+1}}{(2n + 1)!}.$$

(b) Similarly, $\sin z = \cos(z - \pi/2)$, so

$$\sin z = \sum_{n=0}^{\infty}(-1)^n\frac{(z - \pi/2)^{2n}}{(2n)!}.$$

Exercise V.2.8. *Let $f(z) = \frac{1}{(z-1)(z-2)}$. Find the Laurent series for f:*
(a) In the disc $|z| < 1$.
(b) In the annulus $1 < |z| < 2$.
(c) In the region $2 < |z|$.

Solution. We can write

$$\frac{1}{(z - 1)(z - 2)} = \frac{-1}{z - 1} + \frac{1}{z - 2}.$$

If $|z| < 1$, we have

$$\frac{-1}{z - 1} = \frac{1}{1 - z} = \sum_{n=0}^{\infty} z^n$$

and

$$\frac{1}{z - 2} = \frac{-1}{2}\frac{1}{1 - (z/2)} = -\frac{1}{2}\sum_{n=0}^{\infty}\left(\frac{z}{2}\right)^n$$

so

$$\frac{1}{(z - 1)(z - 2)} = \sum_{n=0}^{\infty} z^n - \frac{1}{2}\sum_{n=0}^{\infty}\left(\frac{z}{2}\right)^n.$$

If $1 < |z| < 2$, then

$$\frac{-1}{z - 1} = \frac{-1}{z}\frac{1}{1 - (1/z)} = \frac{-1}{z}\sum_{n=0}^{\infty}\frac{1}{z^n}$$

and

$$\frac{1}{z-2} = \frac{-1}{2} \sum_{n=0}^{\infty} \left(\frac{z}{2}\right)^n$$

so

$$\frac{1}{(z-1)(z-2)} = \sum_{n=0}^{\infty} \frac{-1}{z^{n+1}} - \frac{1}{2} \sum_{n=0}^{\infty} \left(\frac{z}{2}\right)^n.$$

If $2 < |z|$, then

$$\frac{1}{z-2} = \frac{1}{z} \frac{1}{1-(2/z)} = \frac{1}{z} \sum_{n=0}^{\infty} \left(\frac{2}{z}\right)^n$$

so

$$\frac{1}{(z-1)(z-2)} = \sum_{n=0}^{\infty} \frac{-1}{z^{n+1}} + \frac{1}{z} \sum_{n=0}^{\infty} \left(\frac{2}{z}\right)^n.$$

Exercise V.2.9. *Find the Laurent series for* $(z+1)/(z-1)$ *in the region (a)* $|z| < 1$; *(b)* $|z| > 1$.

Solution. (a) Let $f(z) = (z + 1)/(z - 1)$ and suppose that $|z| < 1$. Then

$$f(z) = -\frac{z+1}{1-z} = -(z+1)(1 + z + z^2 + \cdots)$$

which gives, after expansion, the desired Laurent expansion for f in the region $|z| < 1$.

(b) For $|z| > 1$ we have

$$f(z) = \frac{1}{z} \frac{z+1}{1-(1/z)} = \left(1 + \frac{1}{z}\right)\left(1 + \frac{1}{z} + \frac{1}{z^2} + \cdots\right).$$

Expanding the above expression gives the desired Laurent expression of f for $|z| > 1$.

Exercise V.2.10. *Find the Laurent series for* $1/z^2(1 - z)$ *in the regions: (a)* $0 < |z| < 1$; *(b)* $|z| > 1$.

Solution. (a) We write

$$f(z) = \frac{1}{z^2} \frac{1}{1-z} = \frac{1}{z^2}(1 + z + z^2 + \cdots) = \frac{1}{z^2} + \frac{1}{z} + 1 + z + z^2 + \cdots.$$

(b) For $|z| > 1$ we simply have

$$f(z) = \frac{-1}{z^3} \frac{1}{1-(1/z)} = \frac{-1}{z^3}\left(1 + \frac{1}{z} + \frac{1}{z^2} + \cdots\right) = -\frac{1}{z^3} - \frac{1}{z^4} - \frac{1}{z^5} - \cdots.$$

Exercise V.2.11. *Find the power series expansion of*

$$f(z) = \frac{1}{1+z^2}$$

around the point $z = 1$, *and find the radius of convergence of this series.*

Solution. We have the factorization $z^2 + 1 = (z + i)(z - i)$, so the singularities of f are at i and $-i$. Since $|1 - i| = |1 + i| = \sqrt{2}$, the radius of convergence of the power series expansion of f at $z = 1$ is $\sqrt{2}$. Using partial fractions we get the expression

$$f(z) = \frac{i}{2}\left[\frac{1}{z+i} - \frac{1}{z-i}\right].$$

Since we want the power series expansion at $z = 1$ we must make the following transformation,

$$f(z) = \frac{i}{2}\left[\frac{1}{i+1}\frac{1}{\left(1 + \frac{z-1}{i+1}\right)} - \frac{1}{1-i}\frac{1}{\left(1 + \frac{z-1}{1-i}\right)}\right]$$

so that

$$f(z) = \sum_{n=0}^{\infty} \frac{i(-1)^n}{2}\left(\frac{1}{(i+1)^n} - \frac{1}{(1-i)^n}\right)(z-1)^n.$$

Exercise V.2.12. *Find the Laurent expansion of*

$$f(z) = \frac{1}{(z-1)^2(z+1)^2}$$

for $1 < |z| < 2$.

Solution. Write

$$\frac{1}{(z-1)^2(z+1)^2} = \frac{1}{(z^2-1)^2} = \frac{1}{z^4}\frac{1}{(1 - 1/z^2)^2},$$

and since $|z| > 1$ we have

$$\frac{1}{1 - 1/z^2} = \sum_{n=0}^{\infty}(1/z^2)^n.$$

Conclude.

Exercise V.2.13. *Obtain the first four terms of the Laurent series expansion of*

$$f(z) = \frac{e^z}{z(z^2 + 1)}$$

valid for $0 < |z| < 1$.

Solution. We write

$$f(z) = \frac{e^z}{z}\frac{1}{1 - (-z)^2} = \frac{1}{z}\left(1 + z + \frac{z^2}{2!} + \frac{z^3}{3!} + \cdots\right)(1 - z^2 + z^4 + \cdots)$$

$$= \frac{1}{z} + 1 - \frac{z}{2} - \frac{5}{6}z^2 + \cdots.$$

Exercise V.2.14. *Assume that f is analytic in the upper half plane, and that f is periodic of period 1. Show that f has an expansion of the form*

$$f = \sum_{-\infty}^{\infty} c_n e^{2\pi i n z},$$

where

$$c_n = \int_0^1 f(x+iy)e^{-2\pi i n(x+iy)}dx,$$

for any value of $y > 0$. [Hint: Show that there is an analytic function f^* on a disc from which the origin is deleted such that

$$f^*(e^{2\pi i z}) = f(z).$$

What is the Laurent series for f^*? Abbreviate $q = e^{2\pi i z}$.]

Solution. Let z_0 be a point in the unit disc minus the origin. Let z be another point on the unit disc minus the origin and let γ be a path joining z_0 and z. Let

$$g_\gamma(z) = \int_{z_0,\gamma}^z \frac{d\zeta}{\zeta} + \log z_0$$

for some branch of the logarithm. For different paths γ joining z_0 and z and different branches of the logarithm, the function g_γ differs by integral multiples of $2\pi i$. Since f is periodic we see that we can define an analytic function on the unit circle minus the origin by

$$f^*(q) = f\left(\frac{1}{2\pi i}(g_\gamma(q))\right).$$

The coefficients in the Laurent expansion of f^* are given by

$$a_n = \frac{1}{2\pi i}\int_{C_r} \frac{f^*(\zeta)}{\zeta^{n+1}}d\zeta = \frac{1}{2\pi}\int_0^{2\pi} \frac{f^*(re^{i\theta})}{r^n e^{ni\theta}}d\theta.$$

Changing variables $2\pi x = \theta$ and writing $r = e^{-2\pi y}$ we find that the above coefficient is

$$a_n = \int_0^1 f^*(e^{-2\pi y}e^{2\pi i x})e^{2\pi n y}e^{-2\pi i n x}dx = \int_0^1 f^*(e^{2\pi i(x+iy)})e^{-2\pi i n(x+iy)}dx.$$

Therefore $f^*(q) = \sum_{-\infty}^{\infty} a_n q^n$. But $q = e^{2\pi i z} = e^{2\pi i(x+iy)}$ and by construction we have

$$f^*(e^{2\pi i(x+iy)}) = f(x+iy)$$

so $f(x+iy) = \sum_{-\infty}^{\infty} c_n e^{2\pi i n(x+iy)}$ where

$$c_n = \int_0^1 f(x+iy)e^{-2\pi i n(x+iy)}dx.$$

Exercise V.2.15. *Assumptions being as in Exercise 14, suppose in addition that there exists $y_0 > 0$ such that $f(z) = f(x + iy)$ is bounded in the domain $y \geq y_0$. Prove that the coefficients c_n are equal to 0 for $n < 0$. Is the converse true? Proof?*

Solution. Suppose that $y > y_0$ and let B be a bound for f in that region. Then we have the estimate

$$|c_n| \leq \int_0^1 |f(x + iy)| \, |e^{-2\pi i n(x+iy)}| \, dx$$

$$\leq \int_0^1 Be^{2\pi ny} dx = Be^{2\pi ny}.$$

If $n < 0$, letting $y \to \infty$ we see that $c_n = 0$, as was to be shown.

Now we show that the converse is true. Suppose that $c_n = 0$ for all $n < 0$. Then using the notation of Exercise 14 we have $a_n = 0$ for all $n < 0$. Therefore f^* is bounded in a neighborhood of the origin, that is there exists $0 < B$ and $0 < C < 1$ such that for all $|q| < C$ we have $|f^*(q)| < B$. But since $|q| = |e^{2\pi iz}| = e^{-2\pi y}$ and $f^*(e^{2\pi iz}) = f(z)$ we conclude that $|f(z)| < B$ whenever $y > \frac{-1}{2\pi} \log C$. This ends the exercise.

V.3   Isolated Singularities

Exercise V.3.1. *Show that the following series define a meromorphic function on $\mathbf{C}$ and determine the set of poles, and their orders.*

(a) $\sum_{n=0}^\infty \frac{(-1)^n}{n!(n+z)}$

(b) $\sum_{n=1}^\infty \frac{1}{z^2+n^2}$

(c) $\sum_{n=1}^\infty \frac{1}{(z+n)^2}$

(d) $\sum_{n=1}^\infty \frac{\sin nz}{n!(z^2+n^2)}$

(e) $\frac{1}{z} + \sum_{n\neq 0, n=-\infty}^\infty \left[\frac{1}{z-n} + \frac{1}{n}\right]$

Solution. (a) Let $f(z) = \sum_{n=0}^\infty \frac{(-1)^n}{n!(n+z)}$. We contend that f is meromorphic on $\mathbf{C}$ with simple poles at the negative integers. Let $R > 0$, and select an integer N such that $N > 2R$. We write

$$f(z) = \sum_{n=0}^N \frac{(-1)^n}{n!(n + z)} + \sum_{n=N+1}^\infty \frac{(-1)^n}{n!(n + z)}.$$

The first sum exhibits the poles of order 1 at the negative integers of absolute value $\leq N$. The second sum defines a holomorphic function for $|z| < R$. Indeed, for $|z| < R$ and $n > N$ we have

$$\left|\frac{(-1)^n}{n!(n + z)}\right| \leq \frac{1}{n!(n - |z|)} \leq \frac{1}{n!R}$$

and $\sum 1/n!$ converges. The above argument is true for all R, so our contention is proved.

(b) Let $f(z) = \sum_{n=1}^{\infty} \frac{1}{z^2+n^2}$ and let $S = \{\ldots, -2i, -i, i, 2i, \ldots\}$. We contend that f is meromorphic on $\mathbb{C}$ with simple poles at points of S. Let $R > 0$, $N > 2R$ and write

$$f(z) = \sum_{n=1}^{N} \frac{1}{z^2+n^2} + \sum_{n=N+1}^{\infty} \frac{1}{z^2+n^2}.$$

Since $z^2+n^2 = (z+in)(z-in)$, the first sum defines a meromorphic function with simple poles $\pm in$ for $n \leq N$. If $|z| < R$ the second sum defines a holomorphic function. Indeed, with $|z| < R$ and $n > N$ we have the estimates

$$\left|\frac{1}{z^2+n^2}\right| \leq \frac{1}{n^2 - |z^2|} \leq \frac{1}{n^2 - R^2}$$

$$= \frac{1}{n^2}\frac{1}{1-(R/n)^2}$$

$$\leq \frac{4}{3n^2}.$$

The series $\sum 1/n^2$ converges so we have proved our contention.

(c) Let $f(z) = \sum_{n=1}^{\infty} \frac{1}{(z+n)^2}$. We contend that f is a meromorphic function on $\mathbb{C}$ with poles of order 2 at the negative integers. Let $R > 0$, $N > 2R$ and write

$$f(z) = \sum_{n=1}^{N} \frac{1}{(z+n)^2} + \sum_{n=N+1}^{\infty} \frac{1}{(z+n)^2}.$$

The first sum exhibits the poles. If $|z| < R$ the infinite sum defines a holomorphic function. Indeed, if $|z| < R$ and $n > N$ we have

$$\left|\frac{1}{(z+n)^2}\right| \leq \frac{1}{(n-|z|)^2} \leq \frac{1}{(n-R)^2},$$

and the denominator satisfies $(n-R)^2 = n^2(1-R/n)^2 \geq n^2(1-1/2)^2$ so

$$\left|\frac{1}{(z+n)^2}\right| \leq \frac{4}{n^2}.$$

Conclude.

(d) Let $f(z) = \sum_{n=1}^{\infty} \frac{\sin nz}{n!(z^2+n^2)}$. We contend that f is a meromorphic function with simple poles at the points of S where S is defined as in (b). Note that the solutions of the equation $\sin nz = 0$ are $\{\ldots, \frac{-2\pi}{n}, \frac{-\pi}{n}, 0, \frac{\pi}{n}, \frac{2\pi}{n}, \ldots\}$, so the set of zeros of $\sin nz$ and $z^2 + n^2$ are disjoint for all $n \geq 1$. Let $R > 0$, $N > 2R$ and write

$$f(z) = \sum_{n=1}^{N} \frac{\sin nz}{n!(z^2+n^2)} + \sum_{n=N+1}^{\infty} \frac{\sin nz}{n!(z^2+n^2)}.$$

The first sum exhibits the poles. If $|z| < R$, we claim that the second sum defines a holomorphic function. Indeed, the power series expansion of the sin implies $|\sin nz| \leq e^{n|z|} \leq e^{nR}$ and the ratio test shows that $e^{nR}/n!$ converges (actually to

0) as $n \to \infty$, so there exists a constant $C > 0$ such that

$$\left| \frac{\sin nz}{n!(z^2 + n^2)} \right| \le C \frac{1}{|(z^2 + n^2)|}$$

for all $n \ge 1$. The estimate given in (b) for the fraction $1/|(z^2 + n^2)|$ concludes the argument. This proves our contentions.

(e) Let $f(z) = \frac{1}{z} + \sum_{n \neq 0, n=-\infty}^{\infty} \left[\frac{1}{z-n} + \frac{1}{n} \right]$. We contend that f is meromorphic on $\mathbf{C}$ with simple poles at the integers. First we have

$$f(z) = \frac{1}{z} + \sum_{n \ge 1} \frac{z}{n(z-n)} + \sum_{m \ge 1} \frac{-z}{m(z+m)}.$$

Now let $R > 0$, $N > 2R$ and write

$$f(z) = \left(\frac{1}{z} + \sum_{n=1}^{N} \frac{z}{n(z-n)} - \frac{z}{n(z+n)} \right) + \left(\sum_{n=N+1}^{\infty} \frac{z}{n(z-n)} - \frac{z}{n(z+n)} \right).$$

The expression in the first parenthesis exhibits the poles, while the expression in the second parenthesis defines a holomorphic function for $|z| < R$. Indeed, the absolute value of the general term of the series is

$$\left| \frac{2z}{z^2 - n^2} \right| \le \frac{2}{n^2} \frac{R}{1 - |z/n|^2}.$$

However,

$$1 - \left| \frac{z}{n} \right|^2 \ge 1 - \frac{|z^2|}{N^2} \ge 1 - \frac{1}{4}$$

whenever $n > N$, so we see that the infinite sum defines a holomorphic function for $|z| < R$. This proves our contention.

Exercise V.3.2. *Show that the function*

$$f(z) = \sum_{n=1}^{\infty} \frac{z^2}{n^2 z^2 + 8}$$

is defined and continuous for the real values of z. Determine the region of the complex plane in which this function is analytic. Determine its poles.

Solution. Let $f_n(z) = \frac{z^2}{n^2 z^2 + 8}$. The for all real z the inequality $n^2 z^2 + 8 \ge n^2 z^2$ holds, thus $0 \le f_n(z) \le 1/n^2$. Now each partial sum $\sum_{k=1}^{n} f_k$ is continuous, and the inequalities show that the sequence of partial sums converges uniformly to f. Hence f is well defined for real z and f is continuous.

Let S denote the set of zeros of $n^2 z^2 + 8$ union the origin, that is

$$S = \{i\sqrt{8}/n : n \in \mathbf{Z}\} \cup \{0\}.$$

We contend that f is analytic on the complement of S. Let $\delta > 0$ be small, and let $R > 0$ be so large that S is contained in the disc of radius R centered at the origin.

Suppose $|z| < R$ and $|z| > \delta$, so that we stay away from the origin. Choose N so large that $8/N^2 < \delta^2/2$. Write

$$f(z) = \sum_{n=1}^{N} \frac{z^2}{n^2 z^2 + 8} + \sum_{n=N+1}^{\infty} \frac{z^2}{n^2 z^2 + 8}$$

The first sum exhibits the poles. The infinite sum defines a holomorphic function on the region $|z| < R$ and $|z| > \delta$. Indeed, for $n > N$ we have

$$\left| \frac{z^2}{n^2 z^2 + 8} \right| \leq \frac{R^2}{n^2 \delta^2 - 8},$$

and the denominator satisfies

$$n^2 \delta^2 - 8 = n^2 \left(\delta^2 - \frac{8}{n^2} \right) \geq n^2 \left(\delta^2 - \frac{\delta^2}{2} \right)$$

thus

$$\left| \frac{z^2}{n^2 z^2 + 8} \right| \leq \frac{2R^2}{n^2 \delta^2}.$$

Since the series $\sum 1/n^2$ converges, the infinite sum defines a holomorphic function in the region $|z| < R$ and $|z| > \delta$. The above results being true for all large R, and all small δ, our contention is proved.

Exercise V.3.3. *Show that the series*

$$\sum_{n=1}^{\infty} \left(\frac{z+i}{z-i} \right)^n$$

defines an analytic function on a disc of radius 1 centered at $-i$.

Solution. Let $0 < s < 1$. Suppose that z belongs to the open disc centered at $-i$ of radius s. Then $z + i \leq s$ and

$$|z - i| \geq | - 2i| - |z + i| \geq 2 - s.$$

Therefore

$$\left| \frac{z+i}{z-i} \right| \leq \frac{s}{2-s}.$$

Since $0 < s < 1$ we have $0 < s/(2 - s) < 1$, and the series

$$\sum_{n \geq 1} \left(\frac{s}{2-s} \right)^n$$

converges. Since s was arbitrary we have shown that the series

$$\sum_{n \geq 1} \left(\frac{z+i}{z-i} \right)^n$$

defines an analytic function on the disc of radius 1 centered at $-i$.

Exercise V.3.4. *Let $\{z_n\}$ be a sequence of distinct complex numbers such that*

$$\sum \frac{1}{|z_n|^3} \text{ converges.}$$

Prove that the series

$$\sum_{n=1}^{\infty} \left(\frac{1}{(z-z_n)^2} - \frac{1}{z_n^2} \right)$$

defines a meromorphic function on **C**. *Where are the poles of this function?*

Solution. Let $f(z) = \sum_{n=1}^{\infty} \left(\frac{1}{(z-z_n)^2} - \frac{1}{z_n^2} \right)$, and let $R > 0$. We now show that f defines a meromorphic function on the open disc $D(0, R)$ or radius R centered at the origin. The hypothesis that $\sum 1/|z_n|^3$ converges implies that $|z_n| \to \infty$ as $n \to \infty$. So there are only finitely many n such that $z_n \in D(0, R)$. Select N so large that for all $n > N$ we have $|z_n| > 2R$. Each $z_n \in D(0, R)$ is a pole of order 2. Write

$$f(z) = \sum_{n=1}^{N} \left(\frac{1}{(z-z_n)^2} - \frac{1}{z_n^2} \right) + \sum_{n=N+1}^{\infty} \left(\frac{1}{(z-z_n)^2} - \frac{1}{z_n^2} \right).$$

The first sum exhibits the poles in $D(0, R)$, so it is sufficient to show that the second sum defines a holomorphic function on $D(0, R)$. We have the estimates

$$\left| \frac{1}{(z-z_n)^2} - \frac{1}{z_n^2} \right| = \left| \frac{-z^2 + 2zz_n}{z_n^2(z-z_n)^2} \right| = \frac{1}{|z_n|^3} \frac{\left| -\frac{z^2}{z_n} + 2z \right|}{\left| \frac{z}{z_n} - 1 \right|^2}$$

$$\leq \frac{1}{|z_n|^3} \frac{\frac{R^2}{2R} + 2R}{\left(1 - \frac{R}{2R}\right)}$$

$$\leq \frac{B}{|z_n|^3}$$

where B is some large positive constant. By hypothesis $\sum 1/|z_n|^3$ converges, so this completes the proof.

Exercise V.3.5. *Let f be meromorphic on* **C** *but not entire. Let $g(z) = e^{f(z)}$. Show that g is not meromorphic on* **C**.

Solution. Since f is not entire it has at least one pole, say at z_0. In a neighborhood of z_0 we can write

$$(z - z_0)^m f(z) = p(z) + (z - z_0)^m h(z)$$

where h is holomorphic and p is a polynomial of degree $< m$. So

$$f(z) = \frac{p(z)}{(z - z_0)^m} + h(z),$$

and therefore

$$e^{f(z)} = e^{\frac{p(z)}{(z-z_0)^m}} e^{h(z)}.$$

But $e^{h(z)}$ is holomorphic at z_0, and the power series expansion of the exponential shows that $e^{\frac{p(z)}{(z-z_0)^m}}$ has an essential singularity at z_0, so $e^{f(z)}$ is not meromorphic on **C**.

Exercise V.3.6. *Let f be a nonconstant entire function, i.e., a function analytic on all of* **C**. *Show that the image of f is dense in* **C**.

Solution 1. In the spirit of this chapter we use the Casoratti–Weierstrass theorem. Suppose f is an entire function whose image is not dense in **C**. Then there exists a complex number α and a positive number s such that $|f(z)-\alpha| > s$ for all complex z. Write $f(z) = \sum_{n=0}^{\infty} a_n z^n$ and suppose that there are infinitely many nonzero terms in this expansion. Then, for all $z \neq 0$ let $g(z) = f(1/z)$. We see that g has an essential singularity at 0 so by the Casoratti–Weierstrass theorem for some z near 0 we have $|g(z)-\alpha| < s$, hence $|f(1/z)-\alpha| < s$. This contradiction implies that the power series expansion of f can have only finitely many terms. Then the fundamental theorem of algebra guarantees that f is constant. This contradicts the hypothesis.

Solution 2. Suppose there exists a complex number α and a positive number s such that $|f(z) - \alpha| > s$, for all complex z. Then the function $g(z) = 1/(f(z) - \alpha)$ is entire and bounded, so by Liouville's theorem, g is constant. Hence f is constant, again contradicting the hypothesis.

Exercise V.3.7. *Let f be meromorphic on an open set U. Let*

$$\varphi : V \to U$$

be an analytic isomorphism. Suppose that $\varphi(z_0) = w_0$, and f has order n at w_0. Show that $f \circ \varphi$ has order n at z_0. In other words, the order is invariant under analytic isomorphisms. [Here n is a positive or negative integer.]

Solution. We can write

$$f(w) = \sum_{m=n}^{\infty} a_m (w - w_0)^m$$

for all w near w_0. Also we have $\varphi(z) = w_0 + b_1(z - z_0) + b_2(z - z_0)^2 + \cdots$ for all z near z_0. We assume that φ is an analytic isomorphism so $b_1 \neq 0$. Put $w = \varphi(z)$ in the expression of f at w_0. Since $b_1 \neq 0$ we see that the composite $f \circ \varphi$ also has order n at z_0.

Exercise V.3.8. *A meromorphic function f is said to be **periodic** with period w if $f(z + w) = f(z)$ for all $z \in$ **C**. Let f be a meromorphic function, and suppose f is periodic with three periods w_1, w_2, w_3 which are linearly independent over the rational numbers. Prove that f is constant. [Hint: Prove that there exist elements w which are integral linear combinations of w_1, w_2, w_3 and arbitrarily small in absolute value.] The exponential function is an example of a singly periodic function. Examples of doubly periodic functions will be given in Chapter XIV.*

Solution.
Following the hint, we prove that there exists integral linear combinations of w_1, w_2 and w_3 which are arbitrarily close to 0 (and not 0 since the three periods are linearly independent over the rational numbers.) Let $\tilde{w}_1 = w_1/w_3$ and $\tilde{w}_2 = w_2/w_3$. It suffices to show that for every positive integer N, we can find integers m_1, m_2 and n such that

$$|m_1\tilde{w}_1 + m_2\tilde{w}_2 - n| \leq c/N, \quad \text{for some constant } c.$$

We use a modification of the pigeon hole principle (see figure on page 73). Fix N, and choose a positive integer B such that:

$$|\text{Im}(\tilde{w}_i)| < B \quad \text{for} \quad i = 1, 2.$$

Finally, choose an integer M such that $M > 4BN^2$. Consider the box:

$$R = \{z \in \mathbf{C} : 0 \leq \text{Re}(z) < 1 \quad \text{and} \quad |\text{Im}(z)| \leq 2BM\}.$$

For each pair of integers (m_1, m_2), let $\{m_1\tilde{w}_1 + m_2\tilde{w}_2\}$ denote the representative of $m_1\tilde{w}_1 + m_2\tilde{w}_2 \bmod \mathbf{Z}$ with real part between 0 and 1. If $0 < m_1, m_2 \leq M$, then

$$0 \leq \text{Re}(\{m_1\tilde{w}_1 + m_2\tilde{w}_2\}) < 1 \quad \text{and}$$
$$\text{Im}(\{m_1\tilde{w}_1 + m_2\tilde{w}_2\}) = m_1\text{Im}(\tilde{w}_1) + m_2\text{Im}(\tilde{w}_2).$$

so $\{m_1\tilde{w}_1 + m_2\tilde{w}_2\} \in R$. Also, if $(m_1, m_2) \neq (m_1', m_2')$, then

$$\{m_1\tilde{w}_1 + m_2\tilde{w}_2\} \neq \{m_1'\tilde{w}_1 + m_2'\tilde{w}_2\}$$

because w_1, w_2 and w_3 are linearly independent over the rationals. For $0 < m_1$, $m_2 \leq M$, we get M^2 distinct points in R. But we can write R as a union of squares with side $1/N$. There are $4BMN^2$ such squares in R. By assumption $M^2 > 4BMN^2$, so two points $\{m_1'\tilde{w}_1 + m_2'\tilde{w}_2\}$ and $\{m_1''\tilde{w}_1 + m_2''\tilde{w}_2\}$ belong to the same small square. This proves that there exists integers m_1, m_2 and n such that $|m_1\tilde{w}_1 + m_2\tilde{w}_2 - n| \leq \sqrt{2}/N$. This proves that there exists integral linear combinations of w_1, w_2 and w_3 which are arbitrarily close to 0.

Let z_0 be a point where f is holomorphic. Then $f(z_0+w) = f(z_0)$ for arbitrarily small $w \neq 0$, so f is constant in a neighborhood of z_0. Since the set of poles of f is discrete with no points of accumulation, we conclude that f is constant outside of the set of poles. By the theorem on removable singularities we conclude that f is constant on $\mathbf{C}$.

Exercise V.3.9. *Let f be meromorphic on $\mathbf{C}$, and suppose*

$$\lim_{|z|\to\infty} |f(z)| = \infty.$$

Prove that f is a rational function. (You cannot assume as given that f has only a finite number of poles.)

Solution. Since $\lim_{|z|\to\infty} |f(z)| = \infty$, the quantity $|f(z)|$ is well defined for all large $|z|$. This implies that the poles of f are contained in some disc of large radius. Hence f can only have a finite number of poles. Let P be a polynomial

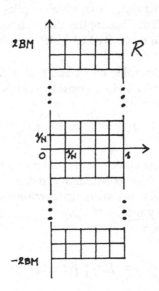

with zeros where f has poles and such that $f_1(z) = f(z)P(z)$ is entire. Let $g(z) = f_1(1/z)$ defined for $z \neq 0$. If f_1 is not a polynomial, then g has an essential singularity at 0. However, $|g(z)| \to \infty$ as $|z| \to 0$, which contradicts the theorem of Casoratti–Weierstrass. So f_1 must be a polynomial, and we are done.

Exercise V.3.10 (The Riemann Sphere).   *Let S be the union of $\mathbf{C}$ and a single point denoted by ∞, and called **infinity**. Let f be a function on S. Let $t = 1/z$, and define*

$$g(t) = f(1/t)$$

*for $t \neq 0, \infty$. We say that f has an **isolated singularity** (resp. is **meromorphic** resp. is **holomorphic**) at infinity if g has an isolated singularity (resp. is meromorphic resp. is holomorphic) at 0. The order of g at 0 will be called the order **order of** f **at infinity**. If g has a removable singularity at 0, and so can be defined as a holomorphic function in a neighborhood of 0, then we say that f is **holomorphic at infinity**.*

We say that f is meromorphic on S, if f is meromorphic on $\mathbf{C}$ and is also meromorphic at infinity. We say that f is holomorphic on S if f is holomorphic on $\mathbf{C}$ and is also holomorphic on at infinity.

Prove:

The only meromorphic functions on S are the rational functions, that is, the quotients of polynomials. The only holomorphic function on S are the constants. If f is holomorphic on $\mathbf{C}$ and has a pole at infinity, then f is a polynomial.

In this last case, how would you describe the order of f at infinity in terms of the polynomial?

Solution. We prove the last two assertions first. Suppose f is holomorphic on S. Then g has a removable singularity at 0, which implies that in a neighborhood of 0, the function g is bounded. This implies that f is bounded outside some disc and therefore f is bounded on $\mathbf{C}$. By Liouville's theorem we conclude that f is constant.

Now suppose f is holomorphic on $\mathbf{C}$, and that f has a pole at infinity. The function f has a certain power series expansion $f(z) = \sum_{n=0}^{\infty} a_n z^n$, and therefore

$$g(t) = \sum_{n=0}^{\infty} a_n \left(\frac{1}{t}\right)^n.$$

Since g has a pole at 0, we conclude that the power series of f at 0 has only finitely many nonzero terms, thus f is a polynomial and $\deg f = -\mathrm{ord}_0\, g$.

Finally, suppose that f is a meromorphic function on S. Since S is compact, f can have only finitely many poles in $\mathbf{C}$, say $\{(z_i, n_i)\}$, $i = 1, \ldots, m$ where n_i is the order of the pole z_i. Then

$$\varphi = f(z) \prod_{i=1}^{M} (z - z_i)^{n_i}$$

is holomorphic on $\mathbf{C}$, and φ is either holomorphic on S or has a pole at infinity. In all cases, we conclude that φ is a polynomial, hence f is a rational function as was to be shown.

Exercise V.3.11. *Let f be a meromorphic function on the Riemann sphere, so a rational function by Exercise 8. Prove that*

$$\sum_P \mathrm{ord}_P\, f = 0,$$

where the sum is taken over all points P which are either points of $\mathbf{C}$, or $P = \infty$.

Solution. In this exercise, all sums and products are finite. By the previous exercise, we know that f is a rational function. We write

$$f(z) = \frac{P_1(z)}{P_2(z)} = K \frac{\prod_j (z - w_j)^{m_j}}{\prod_i (z - z_i)^{n_i}}.$$

The numerator describes the zeros of f, and the denominator describes the poles of f. So

$$\sum_{P \in \mathbf{C}} \mathrm{ord}_P\, f = \sum_j m_j - \sum_i n_i.$$

Now we have to determine the order of f at infinity. We have

$$g(t) = K \frac{\prod_j (\frac{1}{t} - w_j)^{m_j}}{\prod_i (\frac{1}{t} - z_i)^{n_i}} = K \frac{\prod_j t^{-m_j}(1 - tw_j)^{m_j}}{\prod_i t^{-n_i}(1 - tz_i)^{n_j}}$$

and therefore the order of g at 0 is $-\sum_j m_j + \sum_i n_i$. Hence

$$\sum_P \operatorname{ord}_P f = \sum_j m_j - \sum_i n_i - \sum_j m_j + \sum_i n_i = 0$$

as was to be shown.

Exercise V.3.12. *Let $P_i(i = 1, \ldots, r)$ be points of $\mathbf{C}$ or ∞, and let m_i be integers such that*

$$\sum_{i=1}^{r} m_i = 0.$$

Prove that there exists a meromorphic function f on the Riemann sphere such that

$$\operatorname{ord}_{P_i} f = m_i \qquad i = 1, \ldots, r$$

and $\operatorname{ord}_P f = 0$ *if* $P \neq P_i$.

Solution. We may assume without loss of generality that the points P_i are pairwise disjoint and that $m_i \neq 0$ for all i. Suppose first that $P_i \neq \infty$ for all i. Let z_i be the complex number which determines P_i. Then

$$f(z) = \prod_{i=1}^{r}(z - z_i)^{m_i}$$

is meromorphic on S and satisfies the desired property. Indeed, if $P = P_i$, then $\operatorname{ord}_P f = m_i$, if $P \neq P_i$ for all i, and $P \in \mathbf{C}$, then $\operatorname{ord}_P f = 0$. Finally, if $P = \infty$ the previous exercise implies that $\operatorname{ord}_P f = -\sum m_i = 0$.

Now suppose without loss of generality that $P_1 = \infty$. Let

$$f(z) = \prod_{i=2}^{r}(z - z_i)^{m_i}.$$

Then if i is not equal to 1 and $P = P_i$, we have $\operatorname{ord}_P f = m_i$. If $P \neq P_i$ for all i, and $P \in \mathbf{C}$, then $\operatorname{ord}_P f = 0$. Finally, if $P = P_1$ we have

$$\operatorname{ord}_P f = -\sum_{i \neq 1} m_i = m_1.$$

This concludes the exercise.

VI
Calculus of Residues

VI.1   The Residue Formula

Exercise VI.1.1. *Find the residue of the following function at 0:* $(z^2 + 1)/z$.

Solution. $\operatorname{res}_{z=0} f = 1$.

Exercise VI.1.2. *Find the residue of the following function at 0:* $(z^2 + 3z - 5)/z^3$.

Solution. $\operatorname{res}_{z=0} f = 1$.

Exercise VI.1.3. *Find the residue of the following function at 0:*

$$z^3/(z - 1)(z^4 + 2).$$

Solution. $\operatorname{res}_{z=0} f = 0$.

Exercise VI.1.4. *Find the residue of the following function at 0:*

$$(2z + 1)/z(z^3 - 5).$$

Solution. $\operatorname{res}_{z=0} f = -1/5$.

Exercise VI.1.5. *Find the residue of the following function at 0:* $(\sin z)/z^4$.

Solution. $\operatorname{res}_{z=0} f = -1/3!$.

Exercise VI.1.6. *Find the residue of the following function at 0:* $(\sin z)/z^5$.

Solution. $\operatorname{res}_{z=0} f = 0$.

Exercise VI.1.7. *Find the residue of the following function at 0:* $(\sin z)/z^6$.

Solution. $\operatorname{res}_{z=0} f = 1/5!$.

Exercise VI.1.8. *Find the residue of the following function at 0:* $(\sin z)/z^7$.

Solution. $\operatorname{res}_{z=0} f = 0$.

Exercise VI.1.9. *Find the residue of the following function at 0:* e^z/z.

Solution. $\operatorname{res}_{z=0} f = 1$.

Exercise VI.1.10. *Find the residue of the following function at 0:* e^z/z^2.

Solution. $\operatorname{res}_{z=0} f = 1$.

Exercise VI.1.11. *Find the residue of the following function at 0:* e^z/z^3.

Solution. $\operatorname{res}_{z=0} f = 1/2!$.

Exercise VI.1.12. *Find the residue of the following function at 0:* e^z/z^4.

Solution. $\operatorname{res}_{z=0} f = 1/3!$.

Exercise VI.1.13. *Find the residue of the following function at 0:* $z^{-2}\log(1+z)$.

Solution. $\operatorname{res}_{z=0} f = 1$.

Exercise VI.1.14. *Find the residue of the following function at 0:* $e^z/\sin z$.

Solution. $\operatorname{res}_{z=0} f = 1$.

Exercise VI.1.15. *Find the residue of the following function at 1:* $1/(z^2-1)(z+2)$.

Solution. $\operatorname{res}_{z=1} f = 1/6$.

Exercise VI.1.16. *Find the residue of the following function at 1:* $(z^3 - 1)(z + 2)/(z^4 - 1)^2$.

Solution. $\operatorname{res}_{z=1} f = 3^2/4^2$, because

$$\frac{(z^3 - 1)(z + 2)}{(z^4 - 1)^2} = \frac{(z - 1)(1 + z + z^2)(z + 2)}{(z - 1)^2(1 + z + z^2 + z^3)^2}.$$

Exercise VI.1.17. *Factor the polynomial $z^n - 1$ into factors of degree 1. Find the residue at 1 of $1/(z^n - 1)$.*

Solution. If we let $\theta = 2\pi/n$, then

$$z^n - 1 = (z - 1)(z - e^{i\theta})(z - e^{i2\theta})\cdots(z - e^{i(n-1)\theta}).$$

To compute the residue we can differentiate $z^n - 1$ and evaluate at $z = 1$ or note that

$$z^n - 1 = (z - 1)(1 + z + z^2 + \cdots + z^{n-1}),$$

so that $\operatorname{res}_{z=1} 1/(z^n - 1) = 1/n$.

Exercise VI.1.18. *Let $z_1, \ldots, z_n$ be distinct complex numbers. Let C be a circle around z_1 such that C and its interior do not contain z_j for $j > 1$. Let*

$$f(z) = (z - z_1)(z - z_2) \cdots (z - z_n).$$

Find

$$\int_C \frac{1}{f(z)} dz.$$

Solution. Since

$$W(C, z_j) = \begin{cases} 1 & \text{if } j = 1 \\ 0 & \text{if } j > 1 \end{cases}$$

the residue formula implies that the desired integral is equal to $2\pi i \, \mathrm{res}_{z=z_1}(1/f)$. Hence

$$\int_C \frac{1}{f(z)} dz = 2\pi i \prod_{j=2}^{n} \frac{1}{z_1 - z_j}.$$

Exercise VI.1.19. *Find the residue at i of $1/(z^4 - 1)$. Find the integral*

$$\int_C \frac{1}{(z^4 - 1)} dz$$

where C is a circle of radius $1/2$ centered at i.

Solution. We have the factorization $z^4 - 1 = (z - 1)(z + 1)(z - i)(z + i)$, so

$$\mathrm{res}_{z=i} \frac{1}{z^4 - 1} = \frac{-1}{4i}.$$

We could also differentiate $z^4 - 1$ and evaluate at $z = i$. By the residue formula we conclude that

$$\int_C \frac{1}{(z^4 - 1)} dz = 2\pi i \left(\frac{-1}{4i} \right) = \frac{-\pi}{2}.$$

Exercise VI.1.20. *(a) Find the integral*

$$\int_C \frac{1}{z^2 - 3z + 5} dz,$$

where C is a rectangle oriented clockwise, as shown on the figure.

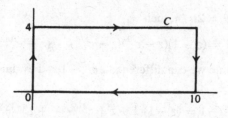

(b) Find the integral $\int_C 1/(z^2 + z + 1)dz$ over the same C.
(c) Find the integral $\int_C 1/(z^2 - z + 1)dz$ over this same C.

Solution. (a) The roots of the polynomial $p(z) = z^2 - 3z + 5$ are

$$z_1 = \frac{3 + i\sqrt{11}}{2} \quad \text{and} \quad z_2 = \frac{3 - i\sqrt{11}}{2}.$$

The only singularity of $1/p(z)$ in the rectangle is at z_1, and to compute the residue we write $p(z) = (z - z_1)(z - z_2)$. Therefore, by the residue formula (and being careful about the orientation of C) we find that

$$\int_C \frac{1}{p(z)}dz = -2\pi i \operatorname{res}_{z=z_1} 1/p(z) = \frac{-2\pi i}{z_1 - z_2} = \frac{-2\pi}{\sqrt{11}}.$$

(b) Let $p(z) = z^2 + z + 1$. Then, the roots of p are

$$z_1 = \frac{-1 + i\sqrt{3}}{2} \quad \text{and} \quad z_2 = \frac{-1 - i\sqrt{3}}{2}.$$

Clearly, none of the roots of p lie in the interior of C, so $1/p$ is holomorphic in this region and therefore

$$\int_C \frac{1}{p(z)}dz = 0.$$

(c) Let $p(z) = z^2 - z + 1$. Then the roots of p are

$$z_1 = \frac{1 + i\sqrt{3}}{2} \quad \text{and} \quad z_2 = \frac{1 - i\sqrt{3}}{2}.$$

The point z_1 is the only singularity of $1/p$ in the interior of C so writing $p(z) = (z - z_1)(z - z_2)$, the residue formula gives

$$\int_C \frac{1}{p(z)}dz = \frac{-2\pi i}{z_1 - z_2} = \frac{-2\pi}{\sqrt{3}}.$$

Exercise VI.1.21. *Let $z_1, \ldots, z_n$ be distinct complex numbers. Determine explicitly the partial fraction decomposition (i.e., the numbers a_i):*

$$\frac{1}{(z - z_1) \cdots (z - z_n)} = \frac{a_1}{z - z_1} + \cdots + \frac{a_n}{z - z_n}.$$

(b) Let $P(z)$ be a polynomial of degree $\leq n - 1$, and let $a_1, \ldots, a_n$ be distinct complex numbers. Assume that there is a partial fraction decomposition of the form

$$\frac{P(z)}{(z - a_1) \cdots (z - a_n)} = \frac{c_1}{z - a_1} + \cdots + \frac{c_n}{z - a_n}.$$

Prove that

$$c_1 = \frac{P(a_1)}{(a_1 - a_2) \cdots (a_1 - a_n)},$$

and similarly for the other coefficients c_j.

Solution. Let

$$f(z) = \frac{1}{(z-z_1)\cdots(z-z_n)} = \frac{a_1}{z-z_1} + \cdots + \frac{a_n}{z-z_n}.$$

These two expressions for f allow us to compute the residue at z_j in two different ways. The first formula gives

$$\text{res}_{z=z_j} f(z) = \prod_{k\neq j} \frac{1}{z_j - z_k},$$

and the second formula gives $\text{res}_{z=z_j} f(z) = a_j$, therefore

$$a_j = \prod_{k\neq j} \frac{1}{z_j - z_k}.$$

(b) Arguing in the exact same way as in (a) we find that

$$c_j = \prod_{k\neq j} \frac{P(a_j)}{(a_j - a_k)}.$$

Exercise VI.1.22. *Let f be analytic on an open disc centered at a point z_0, except at the point itself where f has a simple pole with residue equal to an integer n. Show that there is an analytic function g on the disc such that $f = g'/g$, and*

$$g(z) = (z - z_0)^n h(z) \quad \text{where } h \text{ is analytic and } h(z_0) \neq 0.$$

(To make life simpler, you may assume $z_0 = 0$.)

Solution. Assume $z_0 = 0$, and let $D^* = D - \{0\}$ denote the punctured disc centered at the origin. Let $w \in D^*$ and define

$$g_\gamma(z) = \int_{w,\gamma}^z f(\zeta)d\zeta$$

where the integral is taken on the path $\gamma \subset D^*$ that joins w to z. Suppose we are given two paths γ_1 and γ_2 from w to z. Then the residue formula applied to the closed path $\gamma_1\gamma_2^{-1}$ combined with the fact that f has integer residue at 0 implies that $g_{\gamma_1}(z)$ and $g_{\gamma_2}(z)$ differ by an additive integer multiple of $2\pi i$. Hence

$$g(z) = \exp(g_\gamma(z))$$

is well defined (independent of the path γ) on D^*. Now we show that we can extend g to be analytic at 0. It suffices to show that g is bounded near the origin, so that 0 becomes a removable singularity. Suppose z is close to 0 and let γ be a path joining w and z. If

$$f(z) = \frac{n}{z} + a_0 + a_1 z + \cdots$$

is the Laurent series of f at the origin, then

$$\int_{w,\gamma}^z f(\zeta)d\zeta = \left[\log\gamma(t)^n + a_0\gamma(t) + a_1\frac{\gamma(t)^2}{2} + \cdots\right]_0^1,$$

so $g(z) = z^n h(z)$ where h is analytic on D. Thus g is bounded near the origin and we can extend g to be holomorphic on the whole disc. Therefore $f = g'/g$ on D^* and since f has a simple pole with residue n at 0 we conclude that g has order n at 0, namely there exists an analytic function h such that $g(z) = z^n h(z)$ and $h(0) \neq 0$.

Exercise VI.1.23. *Let f be a function which is analytic on the upper half plane, and on the real line. Assume that there exist numbers $B > 0$ and $c > 0$ such that*

$$|f(\zeta)| \leq \frac{B}{|\zeta|^c}$$

for all ζ. Prove that for any z in the upper half plane, we have the integral formula

$$f(z) = \frac{1}{2\pi i} \int_{-\infty}^{\infty} \frac{f(t)}{t - z} dt.$$

[Hint: Consider the integral over the path shown on the figure, and take the limit as $R \to \infty$.]

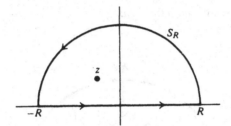

The path consists of a segment from $-R$ to R on the real axis, and the semicircle S_R as shown.

(b) By using a path similar to the previous one, but slightly raised over the real axis, and taking a limit, prove that the formula is still true if instead of assuming that f is analytic on the real line, one merely assumes that f is continuous on the line, but otherwise satisfies the same hypotheses as before.

Solution. Let z be a point in the upper half plane. Choose R so large that z belongs to the interior of S_R. Then by Cauchy's integral formula we have

$$f(z) = \frac{1}{2\pi i} \int_{S_R} \frac{f(\zeta)}{\zeta - z} d\zeta.$$

This formula holds for all large R. Let S_R^+ denote the upper semi circle. Then we can write

$$2\pi i f(z) = \int_{[-R, R]} \frac{f(\zeta)}{\zeta - z} d\zeta + \int_{S_R^+} \frac{f(\zeta)}{\zeta - z} d\zeta.$$

It suffices to show that the last integral goes to 0 as $R \to \infty$. By hypothesis we have the estimate

$$\left| \int_{S_R^+} \frac{f(\zeta)}{\zeta - z} d\zeta \right| \leq \int_{S_R^+} \left| \frac{f(\zeta)}{\zeta - z} \right| d\zeta \leq \int_{S_R^+} \frac{B}{|\zeta|^c |\zeta - z|} d\zeta.$$

For all large R we have $|\zeta - z| \geq R/2$ so

$$\left| \int_{S_R^+} \frac{f(\zeta)}{\zeta - z} d\zeta \right| \leq \frac{2B}{R} \int_{S_R^+} \frac{1}{|\zeta|^c} d\zeta \leq \frac{2B}{R} \frac{\pi R}{R^c} = \frac{2B\pi}{R^c},$$

which proves that the integral over the upper semi circle goes to 0 as $R \to \infty$. Hence

$$2\pi i f(z) = \lim_{R \to \infty} \int_{[-R, R]} \frac{f(\zeta)}{\zeta - z} d\zeta,$$

and therefore

$$f(z) = \frac{1}{2\pi i} \int_{-\infty}^{\infty} \frac{f(t)}{t - z} dt.$$

(b) Consider a path as shown on the figure.

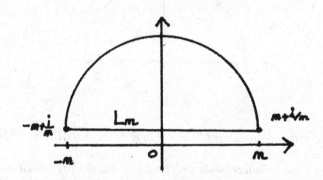

Let L_n be the segment from $n + i/n$ to $-n + i/n$. Then, arguing as before we have

$$2\pi i f(z) = \lim_{n \to \infty} \int_{L_n} \frac{f(\zeta)}{\zeta - z} d\zeta,$$

and

$$\int_{L_n} \frac{f(\zeta)}{\zeta - z} d\zeta = \int_{\mathbf{R}} \chi_{[-n,n]} \frac{f(t + i/n)}{t + i/n - z} dt = \int_{\mathbf{R}} f_n(t) dt$$

where $\chi_{[-n,n]}$ denotes the characteristic function of the interval $[-n, n]$ and

$$f_n(t) = \chi_{[-n,n]} \frac{f(t + i/n)}{t + i/n - z}.$$

By continuity,

$$\lim_{n \to \infty} f_n(t) = \frac{f(t)}{t - z}$$

and the following estimates show that the f_n's are uniformly bounded by an integrable function

$$|f_n(t)| \leq \frac{|f(t + i/n)|}{|t + i/n - z|} \leq \frac{B}{|t + i/n|^c |t + i/n - z|}$$

and $|t + i/n| \geq |t|$. For all large $|t|$ we have

$$|t + i/n - z| \geq \frac{|t|}{2}$$

so that for all large $|t|$ we get

$$|f_n(t)| \leq \frac{K}{|t|^{c+1}}$$

for some positive constant K. We can apply the dominated convergence theorem to obtain

$$f(z) = \frac{1}{2\pi i} \int_{-\infty}^{\infty} \frac{f(t)}{t - z} dt.$$

Exercise VI.1.24. *Determine the poles and find the residues of the following functions.* (a) $1/\sin z$ (b) $1/(1 - e^z)$ (c) $z/(1 - \cos z)$.

Solution. (a) The function $1/\sin z$ has poles at the points $z = k\pi$ where $k \in \mathbf{Z}$, and the derivative of $\sin z$, is $\cos z$ so

$$\operatorname{res}_{z=k\pi} \frac{1}{\sin z} = (-1)^k.$$

(b) The poles of $1/(1 - e^z)$ are the points $2\pi i k$ where $k \in \mathbf{Z}$. The derivative of $1 - e^z$ is $-e^z$ so

$$\operatorname{res}_{z=2\pi i k} \frac{1}{1 - e^z} = -1.$$

(c) The solutions of $1 - \cos z = 0$ are the complex numbers $z = 2\pi k$ with $k \in \mathbf{Z}$, and $z/(1 - \cos z)$ has poles at these points. Moreover

$$\frac{z}{1 - \cos z} = \frac{1}{z/2! - z^3/4! + \cdots},$$

so

$$\operatorname{res}_{z=2\pi k} \frac{z}{1 - \cos z} = \begin{cases} 2 & \text{if } k = 0, \\ 0 & \text{if } k \neq 0. \end{cases}$$

Exercise VI.1.25. *Show that*

$$\int_{|z|=1} \frac{\cos e^{-z}}{z^2} dz = 2\pi i \cdot \sin 1.$$

Solution. Let $f(z) = \cos e^{-z}$. By Cauchy's integral formula we have

$$2\pi i f'(0) = \int_{|z|=1} \frac{\cos e^{-z}}{z^2} dz.$$

Differentiating we obtain $f'(0) = \sin 1$ which proves the desired formula.

Exercise VI.1.26. *Find the integrals, where C is the circle of radius 8 centered at the origin.*
(a) $\int_C \frac{1}{\sin z} dz$ (b) $\int_C \frac{1}{1-\cos z} dz$
(c) $\int_C \frac{1+z}{1-e^z} dz$ (d) $\int_C \tan z\, dz$
(e) $\int_C \frac{1+z}{1-\sin z} dz$

Solution. (a) The function $1/\sin z$ has poles at the numbers πk where $k \in \mathbf{Z}$. The residue at πk is $(-1)^k$ (see Exercise 24), so

$$\int_C \frac{1}{\sin z} dz = 2\pi i.$$

(b) The zeros of $1 - \cos z$ are located at the points $z = 2\pi k$ with $k \in \mathbf{Z}$. We first compute the residue at 0. To do so, we use the power series expansion of the cosine:

$$\frac{1}{1 - \cos z} = \frac{1}{\frac{z^2}{2!} - \frac{z^4}{4!} + \frac{z^6}{6!} - \cdots}$$

$$= \frac{2}{z^2} \frac{1}{1 - \frac{2!z^2}{4!} + \frac{2!z^4}{6!} - \cdots}$$

$$= \frac{2}{z^2} \left(1 + \left(\frac{2!z^2}{4!} - \frac{2!z^4}{6!} + \cdots \right) + \left(\frac{2!z^2}{4!} - \frac{2!z^4}{6!} + \cdots \right)^2 + \cdots \right)$$

and we see that the residue of $1/(1 - \cos z)$ at the origin is 0. By periodicity we conclude that the residues of $1/(1 - \cos z)$ at the points $2\pi k$ are all 0 and therefore

$$\int_C \frac{1}{1 - \cos z} dz = 0.$$

(c) Let $f(z) = (1+z)/(1-e^z)$. This function has simple poles at $2\pi i k$ with $k \in \mathbf{Z}$. Simple computations give

$$\text{res}_{z=0} f = 1, \quad \text{res}_{z=-2\pi i} f = -(1 - 2\pi i) \quad \text{and} \quad \text{res}_{z=2\pi i} f = -(1 + 2\pi i).$$

By the residue formula we have

$$\int_C \frac{1+z}{1-e^z} dz = -6\pi i.$$

(d) Let $I = -\int_C \tan z\, dz$. Then

$$I = \int_C \frac{f'}{f}$$

where $f(z) = \cos z$. By the residue formula,

$$I = 2\pi i \sum (\text{number of zeros of } f \text{ in } C).$$

The zeros of f are at the points $k\pi/2$ with k odd. Therefore, f has 6 zeros in the interior of C which gives

$$\int_C \tan z \, dz = -12\pi i.$$

(e) Let $f(z) = (1 + z)/(1 - \sin z)$. The poles of f are precisely at the points $z = k\pi/2$ where k is an odd integer. Since $\sin z = \cos(z - \pi/2)$ we see that

$$f(z) = \frac{1 + z}{1 - \cos(z - \pi/2)}.$$

Arguing like in (b) we see that the residue of f at the points $k\pi/2$ with k odd is 0 and therefore

$$\int_C \frac{1 + z}{1 - \sin z} dz = 0.$$

Exercise VI.1.27. *Let f be holomorphic on and inside the unit circle, $|z| \leq 1$, except for a pole of order 1 at a point z_0 on the circle. Let $f = \sum a_n z^n$ be the power series for f on the open disc. Prove that*

$$\lim_{n \to \infty} \frac{a_n}{a_{n+1}} = z_0.$$

Solution. For z near z_0 we have

$$f(z) = \frac{c}{z - z_0} + \text{higher terms}.$$

Let

$$g(z) = f(z) - \frac{c}{z - z_0}.$$

Then g is analytic in the closure of the unit disc, and

$$g(z) = \sum a_n z^n + \frac{c}{z_0} \sum \left(\frac{z}{z_0}\right)^n = \sum \left(a_n + \frac{c}{z_0^{n+1}}\right) z^n.$$

The power series of g has radius of convergence > 1 so $a_n + c/z_0^{n+1} \to 0$ and therefore $a_n/a_{n+1} \to z_0$.

Exercise VI.1.28. *Let a be real > 1. Prove that the equation*

$$ze^{a-z} = 1$$

has a single solution with $|z| \leq 1$, which is real and positive.

Solution. Let $f(z) = ze^{a-z}$ and $g(z) = f(z) - 1$. Then $|f(z) - g(z)| = 1$ and if $|z| = 1$ with $z = x + iy$, $x, y \in \mathbf{R}$, then

$$|f(z)| = e^{a-x}.$$

But since $a > 1$ we have $|f(z)| > 1$ whenever $|z| = 1$ and therefore $|f(z)-g(z)| < |f(z)|$ on the circle. By Rouché's theorem we conclude that the equations $ze^{a-z} = 1$ and $ze^{a-z} = 0$ have the same number of solutions in the closed unit disc. Since the second equation has only one solution, we conclude that $ze^{a-z} = 1$ has only one solution in the closed unit disc.

Let $f(x) = xe^{a-x}$. Then $f(0) = 0$ and $f(1) > 1$, so the solution of $ze^{a-z} = 1$ is real and positive.

Exercise VI.1.29. *Let U be a connected open set, and let D be an open disc whose closure is contained in U. Let f be analytic on U and not constant. Assume that the absolute value $|f|$ is constant on the boundary of D. Prove that f has at least one zero in D. [Hint: Consider $g(z) = f(z) - f(z_0)$ with $z_0 \in D$.]*

Solution. Fix any $z_0 \in D$ and let $g(z) = f(z) - f(z_0)$. Clearly, we have

$$|f(z) - g(z)| = |f(z_0)|.$$

The maximum modulus principle applied to f in D, combined with the fact that $|f|$ is constant on the boundary of D and that f is not constant implies

$$|f(z_0)| < |f(z)| \quad \text{for all } z \text{ on the boundary of } D.$$

By Rouché's theorem, f and g have the same number of zeros in D. Clearly, g has at least one zero, namely z_0, so f has at least one zero in D. This concludes the proof.

Exercise VI.1.30. *Let f be a function analytic inside and on the unit circle. Suppose that $|f(z) - z| < |z|$ on the unit circle.*
(a) Show that $|f'(1/2)| \le 8$.
(b) Show that f has precisely one zero inside the unit circle.

Solution. (a) By Cauchy's integral formula

$$f'(1/2) = \frac{1}{2\pi i} \int_C \frac{f(\zeta)}{(\zeta - 1/2)^2} d\zeta$$

where C is the unit circle. But $|\zeta - 1/2| \ge 1/2$ for all $\zeta \in C$ and

$$|f(\zeta)| \le |f(\zeta) - \zeta| + |\zeta| < 2|\zeta|$$

by hypothesis. Putting these two observations together we get

$$|f'(1/2)| \le \frac{1}{2\pi} \frac{2 \times 2\pi}{(1/2)^2} = 8.$$

(b) Rouché's theorem implies at once that f has precisely one zero inside the unit circle.

Exercise VI.1.31. *Determine the number of zeros of the polynomial*

$$z^{87} + 36z^{57} + 71z^4 + z^3 - z + 1$$

inside the circle
(a) of radius 1,

(b) of radius 2, centered at the origin.
(c) Determine the number of zeros of the polynomial

$$2z^5 - 6z^2 + z + 1 = 0$$

in the annulus $1 \le |z| \le 2$.

Solution. (a) Let $g(z)$ be the polynomial given in the exercise, and let $f(x) = 71z^4$. If $|z| = 1$, the triangle inequality implies

$$|g(z) - f(z)| \le 1 + 36 + 1 + 1 + 1 < 71,$$

so on the unit circle we have $|g(z) - f(z)| < |f(z)|$. By Rouché's theorem we conclude that $g(z) = 0$ has 4 zeros inside the unit circle.
(b) Let $g(z)$ be the polynomial given in the exercise and let $f(z) = z^{87}$. Then if $|z| = 2$, we have

$$|g(z) - f(z)| \le 36 \times 2^{57} + 71 \times 2^4 + 2^3 + 2 + 1 \le 2^{87} = |f(z)|,$$

so by Rouché's theorem we conclude that $g(z) = 0$ has 87 zeros inside the circle of radius 2 centered at 0.
(c) Let $g(z) = 2z^5 - 6z^2 + z + 1$, $f_1(z) = -6z^2$ and $f_2(z) = 2z^5$. We denote by D_r the open disc of radius r centered at the origin. Rouché's theorem applied to g and f_2 shows that g has 5 zeros in D_2 and no zero on the boundary of D_2. Applying Rouché's theorem to g and f_1 we find that g has two zeros in D_1 and no zeros on the boundary of D_1. Therefore, g has 3 zeros in the annulus $1 \le |z| \le 2$.

Exercise VI.1.32. *Let f, h be analytic on the closed unit disc of radius R, and assume that $f(z) \ne 0$ for z on the circle of radius R. Prove that there exists $\epsilon > 0$ such that $f(z)$ and $f(z) + \epsilon h(z)$ have the same number of zeros inside the circle or radius R. Loosely speaking, we may say that f and a small perturbation of f have the same number of zeros inside the circle.*

Solution. Let $\epsilon > 0$, and define $g_\epsilon(z) = f(z) - \epsilon h(z)$. Then

$$|g_\epsilon(z) - f(z)| \le \epsilon |h(z)|.$$

There exists $\delta > 0$ such that $|f(z)| > \delta$ for all z on the boundary of C (the circle of radius R) because $|f|$ is continuous on this circle and never 0 by hypothesis. Since h is continuous on the same circle, there exist $\epsilon_0 > 0$ so small that $\epsilon_0 |h(z)| < \delta$ whenever $z \in C$. By construction, we have

$$|g_{\epsilon_0}(z) - f(z)| < |f(z)| \quad \text{for all } z \in C.$$

Rouché's theorem guarantees that ϵ_0 verifies the desired property.

Exercise VI.1.33. *Let $f(z) = a_n z^n + \cdots + a_0$ be a polynomial with $a_n \ne 0$. Use Rouché's theorem to show that $f(z)$ and $a_n z^n$ have the same number of zeros in a disc of radius R for R sufficiently large.*

Solution. Select R_0 so large that

$$|a_n| > \frac{|a_{n-1}|}{R_0} + \cdots + \frac{|a_0|}{R_0^n}.$$

Let $g(z) = a_n z^n$ and suppose that $|z| = R_0$. Then

$$|g(z) - f(z)| \le |a_{n-1}|R_0^{n-1} + \cdots + |a_0| \le |a_n|R_0^n = |g(z)|.$$

By Rouché's theorem we conclude that R_0 satisfies the desired property.

Exercise VI.1.34. *(a) Let f be analytic on the closed unit disc. Assume that $|f(z)| = 1$ if $|z| = 1$, and f is not constant. Prove that the image of f contains the unit disc.*
(b) Let f be analytic on the closed unit disc $\overline{D}$. Assume that there exists some point $z_0 \in D$ such that $|f(z_0)| < 1$, and that $|f(z)| \ge 1$ if $|z| = 1$. Prove that $f(D)$ contains the unit disc.

Solution. (a) It suffices to show that if $|w_0| < 1$, then $g(z) = f(z) - w_0$ has a zero in D, the unit disc. If $|z| = 1$, then

$$|g(z) - f(z)| = |w_0| < 1 = |f(z)|,$$

so by Rouché's theorem f and g have the same number of zeros in D. We have reduced the problem to showing that f has a zero in D.

By the maximum modulus principle and the hypothesis on f we see that there exists $z_0 \in D$ such that $f(z_0) \in D$. Let $g(z) = f(z) - f(z_0)$. Then

$$|g(z) - f(z)| = |f(z_0)| < 1 = |f(z)| \quad \text{whenever } |z| = 1$$

so by Rouché's theorem f and g have the same number of zeros in D. Since g has at least one zero in D we get the desired conclusion.
(b) From the hypothesis given, it is clear that with a slight modification, the proof of part (a) carries over.

Exercise VI.1.35. *Let $P_n(z) = \sum_{k=0}^n z^k / k!$. Given R, prove that P_n has no zeros in the disc of radius R for all n sufficiently large.*

Solution. Let $f(z) = e^z$. We know that f has no zeros in C, so $|f(z)| \ge \delta_R > 0$ on the closure $\overline{D}_R$ of the disc or radius R centered at the origin. Moreover, $P_n \to f$ uniformly on $\overline{D}_R$, so for all sufficiently large n, P_n has not zeros in $\overline{D}_R$.

Exercise VI.1.36. *Let $z_1, \ldots, z_n$ be distinct complex numbers contained in the disc $|z| < R$. Let f be analytic on the closed disc $\overline{D}(0, R)$. Let*

$$Q(z) = (z - z_1) \cdots (z - z_n).$$

Prove that

$$P(z) = \frac{1}{2\pi i} \int_{C_R} f(\zeta) \frac{1 - Q(z)/Q(\zeta)}{\zeta - z} d\zeta$$

is a polynomial of degreen -1 having the same value as f at the points $z_1, \ldots, z_n$.

Solution. Since $Q(z_k) = 0$ for all $k = 1, \ldots, n$ we see from Cauchy's integral formula that

$$P(z_k) = \frac{1}{2\pi i} \int_{C_R} f(\zeta) \frac{f(\zeta)}{\zeta - z} d\zeta = f(z_k).$$

To see why P is a polynomial of degree $n - 1$ we write

$$P(z) = \frac{1}{2\pi i} \int_{C_R} \frac{f(\zeta)}{Q(\zeta)} \frac{Q(\zeta) - Q(z)}{\zeta - z} d\zeta.$$

Viewed as a polynomial in z, $Q(\zeta) - Q(z)$ has degree n so there exists $a_k(\zeta)$ such that

$$Q(\zeta) - Q(z) = (\zeta - z) \sum_{k=0}^{n-1} a_k(\zeta) z^k,$$

thus

$$P(z) = \sum_{k=0}^{n-1} \left(\frac{1}{2\pi i} \int_{C_R} \frac{f(\zeta)}{Q(\zeta)} a_k(\zeta) d\zeta \right) z^k.$$

Exercise VI.1.37. *Let f be analytic on $\mathbf{C}$ with the exception of a finite number of isolated singularities which may be poles. Define the **residue at infinity***

$$\operatorname{res}_\infty f(z) dz = -\frac{1}{2\pi i} \int_{|z|=R} f(z) dz$$

for R so large that f has no singularities in $|z| \geq R$.
(a) Show that $\operatorname{res}_\infty f(z) dz$ is independent of R.
(b) Show that the sum of the residues of f at all singularities and the residue at infinity is equal to 0.

Solution. (a) The residue at infinity of f is independent of R because f is holomorphic outside some large disc and since any two circles of large radius are homotopic, this implies that the integrals of f along these two circles are equal.
(b) Suppose that $z_1, \ldots, z_n$ are the poles of f with residues $c_1, \ldots, c_n$ respectively. Then for all sufficiently large R, the residue formula gives

$$\int_{|z|=R} f(\zeta) d\zeta = 2\pi i \sum c_k.$$

It is now clear that

$$\operatorname{res}_\infty f(z) dz + \sum_k \operatorname{res}_{z_k} f(z) = 0.$$

Exercise VI.1.38 (Cauchy's Residue Formula on the Riemann Sphere). *Recall Exercise 2 of Chapter V, §3 on the Riemann sphere. By a (meromorphic) **differential** ω on the Riemann sphere S, we mean an expression of the form*

$$\omega = f(z) dz,$$

*where f is a rational function. For any point $z_0 \in \mathbf{C}$ the **residue** of ω at z_0 is defined to be the usual residue of $f(z) dz$ at z_0. For the point ∞, we write $t = 1/z$,*

$$dt = -\frac{1}{z^2} dz \quad and \quad dz = -\frac{1}{t^2} dt,$$

so we write

$$\omega = f(1/t)\left(-\frac{1}{t^2}\right) dt = -\frac{1}{t^2} f(1/t) dt.$$

*The **residue** of ω at **infinity** is then defined to be the residue of $-\frac{1}{t^2} f(1/t) dt$ at $t = 0$. Prove:*
(a) $\sum$ residues $\omega = 0$ if the sum is taken over all points of $\mathbf{C}$ and also infinity.
(b) Let γ be a circle of radius R centered at the origin in $\mathbf{C}$. If R is sufficiently large, show that

$$\frac{1}{2\pi i} \int_\gamma f(z) dz = -\text{residue of } f(z) dz \text{ at infinity.}$$

(Instead of a circle, you can also take a simple closed curve such that all the poles of f in $\mathbf{C}$ lie in its interior.)
(c) If R is arbitrary, and f has no poles on the circle, show that

$$\frac{1}{2\pi i} \int_\gamma f(z) dz$$
$$= -\sum \text{residues of } f(z) dz \text{ outside the circle, including the residue at } \infty.$$

[Note: In dealing with (a) and (b), you can either find a direct algebraic proof of (a), as in Exercise 38 and deduce (b) from it, or you can prove (b) directly, using a change of variables $t = 1/z$, and the deduce (a) from (b). You probably should carry both ideas out completely to understand fully what's going on.]

Solution. (b) Choose R so large that f has all its singularities and zeros in $D(0, R)$. Then we can make the change of variable $t = 1/z$ in the integral $\frac{1}{2\pi i} \int_\gamma f(z) dz$ and we obtain

$$-\frac{1}{2\pi i} \int_\gamma \frac{-1}{t^2} f(1/t) dt.$$

The first minus signs comes from the fact that the change of variables reverses the orientation. So we have proved (Residue formula) that

$$\frac{1}{2\pi i} \int_\gamma f(z) dz = -\text{residue of } f(z) dz \text{ at infinity.}$$

(a) When R is large, the integral $\frac{1}{2\pi i} \int_{C_R} f(z) dz$ is equal to the sum of the residues of f at points in $\mathbf{C}$. By (b) it follows that $\sum$ residues $\omega = 0$. See the next exercise for a direct (algebraic) proof of (a). It is clear that (b) follows immediately from (a) by the residue formula.
(c) By the residue formula, the integral $\frac{1}{2\pi i} \int_\gamma f(z) dz$ is precisely equal to the sum of the residues of points in the circle of radius R, so part (a) implies at once the desired formula.

Exercise VI.1.39. *(a) Let $P(z)$ be a polynomial. Show directly from the power series expansion of $P(z) dz$ that $P(z)$ has 0 residue in $\mathbf{C}$ and at infinity.*
(b) Let α be a complex number. Show that $1/(z - \alpha)$ has residue -1 at infinity.

(c) Let m be an integer ≥ 2. Show that $1/(z - \alpha)^m$ has residue 0 at infinity and at all complex numbers.

*(d) Let $f(z)$ be a rational function. The theorem concerning the **partial fraction decomposition** of f states that f has an expression*

$$f(z) = \sum_{i=1}^{r} \sum_{m=1}^{n_i} \frac{a_{im}}{(z - \alpha_i)^m} + P(z)$$

where $\alpha_1, \ldots, \alpha_r$ are the roots of the denominator of f, a_{im} are constants, and P is some polynomial. Using this theorem, give a direct (algebraic) proof of Exercise 38(a).

Solution. (a) Since P is entire, the residue of $P(z)dz$ at any points in the complex plane is 0. If we write $P(z) = a_0 + \cdots + a_n z^n$, then

$$\frac{-1}{t^2} P\left(\frac{1}{t}\right) = \frac{-1}{t^2} \left(a_0 + \frac{a_1}{t} + \cdots + \frac{a_n}{t^n}\right).$$

We see from this expression that the residue of $P(z)dz$ at infinity is 0.
(b) Immediate from the expression

$$\frac{-1}{t^2} \frac{1}{(1/t) - \alpha} = \frac{-1}{t(1 - t\alpha)}.$$

(c) The only singularity of the function $1/(z - \alpha)^m$ in $\mathbf{C}$ is at α where this function has residue 0. At infinity, the residue is also 0 because

$$\frac{-1}{t^2} \frac{1}{(1/t - \alpha)^m} = \frac{-t^{m-2}}{(1 - t\alpha)^m}$$

and because we assumed $m \geq 2$.
(d) We prove that the sum of the residues is equal to 0. We use (a), (b), (c), the expression of f, and the fact that the residue of the function in (b) is 1 at α and 0 everywhere else in $\mathbf{C}$. By (b) we are only interested in the terms with $m = 1$, hence the sum of the residues of f taken over the points in $\mathbf{C}$ is

$$a_{11} + a_{21} + \cdots + a_{r1}.$$

Also, by all the previous results we find that the sum of the residues at infinity are

$$-a_{11} - a_{21} - \cdots - a_{r1}.$$

The formula $\sum$ residues ω drops out.

Exercise VI.1.40. *Let $a, b \in \mathbf{C}$ with $|a|$ and $|b| < R$. Let C_R be the circle of radius R. Evaluate*

$$\int_{C_R} \frac{z\,dz}{\sqrt{(z - a)(z - b)}}.$$

The square root is chosen so that the integrand is continuous for $|z| > R$ and has limit 1 as $|z| \to \infty$.

Solution. Consider a circle centered at the origin, of radius $r > R$. Then

$$\int_{C_r} \frac{z\,dz}{\sqrt{(z-a)(z-b)}} = \int_{C_R} \frac{z\,dz}{\sqrt{(z-a)(z-b)}},$$

because the two circles C_r and C_R are homotopic in an open set where the integrand is holomorphic. But

$$\int_{C_r} \frac{z\,dz}{\sqrt{(z-a)(z-b)}} = \int_0^{2\pi} \frac{re^{i\theta}}{\sqrt{(re^{i\theta}-a)(re^{i\theta}-b)}} rie^{i\theta}\,d\theta.$$

We know that the fraction will tend to 1 as $r \to \infty$, but since this is multiplied by $rie^{i\theta}$ we cannot conclude. Therefore, we add and subtract 1

$$\int_0^{2\pi} \frac{re^{i\theta}}{\sqrt{(re^{i\theta}-a)(re^{i\theta}-b)}} rie^{i\theta}\,d\theta$$

$$= \int_0^{2\pi} \left(\frac{re^{i\theta}}{\sqrt{(re^{i\theta}-a)(re^{i\theta}-b)}} - 1 \right) rie^{i\theta} + rie^{i\theta}\,d\theta$$

$$= \int_0^{2\pi} \left(\frac{re^{i\theta}}{\sqrt{(re^{i\theta}-a)(re^{i\theta}-b)}} - 1 \right) rie^{i\theta}\,d\theta.$$

But

$$\frac{re^{i\theta}}{\sqrt{(re^{i\theta}-a)(re^{i\theta}-b)}} - 1$$

$$= \frac{re^{i\theta} - \sqrt{(re^{i\theta}-a)(re^{i\theta}-b)}}{\sqrt{(re^{i\theta}-a)(re^{i\theta}-b)}}$$

$$= \frac{r^2 e^{2i\theta} - (re^{i\theta}-a)(re^{i\theta}-b)}{(\sqrt{(re^{i\theta}-a)(re^{i\theta}-b)})(re^{i\theta} + \sqrt{(re^{i\theta}-a)(re^{i\theta}-b)})}$$

$$= \frac{(a+b)re^{i\theta} - ab}{(\sqrt{(re^{i\theta}-a)(re^{i\theta}-b)})(re^{i\theta} + \sqrt{(re^{i\theta}-a)(re^{i\theta}-b)})}$$

so if we denote by $D(z)$ the denominator, we find that the integrand we are interested in is

$$\frac{(a+b)re^{i\theta}rie^{i\theta}}{D(z)} - \frac{abrie^{i\theta}}{D(z)}.$$

As $|z| = r \to \infty$, the term on the right behaves like $\frac{-abiz}{z(z+z)}$ and the term on the left behaves like $\frac{(a+b)ziz}{z(z+z)} = \frac{(a+b)i}{2}$. An application of the dominated convergence theorem shows that the integral of the term on the right goes to 0, and the integral of the term on the left goes to

$$2\pi \frac{(a+b)i}{2} = \pi(a+b)i.$$

Hence

$$\int_{C_R} \frac{z\,dz}{\sqrt{(z-a)(z-b)}} = \pi(a+b)i.$$

VI.2   Evaluation of Definite Integrals

Exercise VI.2.1. *Find the following integrals:*
(a) $\int_{-\infty}^{\infty} \frac{1}{x^6+1}\,dx = 2\pi/3.$
(b) Show that for a positive integer $n \geq 2$,

$$\int_0^\infty \frac{1}{1+x^n}\,dx = \frac{\pi/n}{\sin \pi/n}.$$

[Hint: Try the path from 0 to R, then from R to $Re^{2\pi i/n}$, then back to 0, or apply a general theorem.]

Solution. (a) Consider the contour shown on the figure, namely a symmetric segment on the real line and a semicircle in the upper half plane.

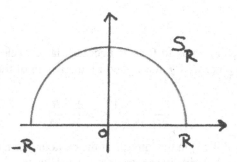

We have

$$\left| \int_{S_R} \frac{1}{1+z^6}\,dz \right| \leq \pi R \frac{B}{R^6}$$

for some constant B valid for all large R. This shows that the integral on the semicircle goes to zero as R tends to infinity, and by the residue formula

$$\int_{-\infty}^{\infty} \frac{1}{x^6+1}\,dx = 2\pi i \sum \text{ residues of } \frac{1}{1+z^6} \text{ in the upper half plane.}$$

The poles of $1/(1+z^6)$ in the upper half plane are at the points $e^{i\pi/6}$, $e^{i\pi/2}$ and $e^{i5\pi/6}$. Moreover, these poles are simple, so we can use the derivative to find the residues. It follows that the desired integral is

$$\int_{-\infty}^{\infty} \frac{1}{1+x^6} dx = 2\pi i \left(\frac{e^{-5i\pi/6}}{6} + \frac{e^{-5i\pi/2}}{6} + \frac{e^{-25i\pi/2}}{6} \right)$$

$$= \frac{\pi i}{3} \left(-\frac{\sqrt{3}}{2} - \frac{i}{2} - i - \frac{i}{2} + \frac{\sqrt{3}}{2} \right) = \frac{2\pi}{3}.$$

(b) We split the contour integral given in the hint in three parts, L_R the segment from 0 to R, A_R the arc from R to $Re^{2\pi i/n}$, and L'_R the segment from $Re^{2\pi i/n}$ to 0.

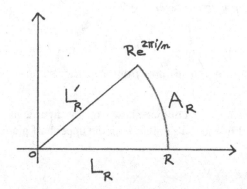

The integral on the arc tends to 0 as R becomes large because this integral is estimated by the sup norm of f multiplied by the length of the arc, and because we assume $n \geq 2$

$$\left| \int_{A_R} \frac{1}{1+z^n} dz \right| \leq R \frac{2\pi}{n} \frac{B}{R^n}.$$

The only pole of $1/(1+z^n)$ in the interior of the contour (for large R) is $e^{\pi i/n}$ and this pole is simple. The derivative shows that the residue is

$$\frac{1}{n} e^{-(n-1)\pi i/n} = \frac{-1}{n} e^{\pi i/n}.$$

Parametrizing L'_R by $te^{2\pi i/n}$ with $0 \leq t \leq R$ we find that

$$\int_{L'_R} \frac{1}{1+z^n} dz = -e^{2\pi i/n} \int_{L_R} \frac{1}{1+z^n} dz.$$

Taking the limit as $R \to \infty$ and using the residue formula we get

$$(1 - e^{2\pi i/n}) \int_0^\infty \frac{1}{1+x^n} dx = 2\pi i \left(\frac{-1}{n} e^{\pi i/n} \right),$$

thus

$$\frac{(e^{\pi i/n} - e^{-\pi i/n})}{2i} \int_0^\infty \frac{1}{1+x^n} dx = \pi/n.$$

By Euler's formula we conclude that

$$\int_0^\infty \frac{1}{1+x^n} dx = \frac{\pi/n}{\sin \pi/n}.$$

Exercise VI.2.2. *Find the following integrals:*

(a) $\int_{-\infty}^\infty \frac{x^2}{x^4+1} dx = \pi\sqrt{2}/2$.

(b) $\int_0^\infty \frac{x^2}{x^6+1} dx = \pi/6$.

Solution. (a) Let $f(z) = z^2/(1+z^4)$. To use the contour given in the text, i.e., a segment on the real line and a semicircle in the upper half plane (see the first figure of the preceding exercise) we must show that f decreases rapidly at infinity. There exists a constant B such that for all large R we have

$$|f(z)| \le B\frac{R^2}{R^4} = \frac{B}{R^2} \quad \text{whenever } |z| = R.$$

The integral on the semicircle is estimated by the sup norm of f multiplied by the length of the semicircle. Hence the integral on the semicircle is bounded by $\pi R(B/R^2) = \pi B/R$, and therefore this integral tends to 0 as R tends to infinity. So

$$\int_{-\infty}^\infty f(x)dx = 2\pi i \sum \text{residues of } \frac{z^2}{1+z^4} \text{ in the upper half plane.}$$

The function $f(z)$ has two simple poles in the upper half plane at $e^{\pi i/4}$ and $e^{3\pi i/4}$. Using the derivative of the denominator and the fact that the numerator is entire, we find that the residues are

$$\frac{(e^{\pi i/4})^2}{4e^{3\pi i/4}} \quad \text{and} \quad \frac{(e^{3\pi i/4})^2}{4e^{9\pi i/4}},$$

respectively. Hence

$$\int_{-\infty}^\infty f(x)dx = 2\pi i \left(\frac{e^{\pi i/2}}{4e^{3\pi i/4}} + \frac{e^{3\pi i/2}}{4e^{\pi i/4}} \right)$$

$$= \frac{\pi i}{2}(e^{-\pi i/4} + e^{5\pi i/4})$$

$$= \frac{\pi i}{2}\left(-2i\frac{\sqrt{2}}{2} \right) = \frac{\pi\sqrt{2}}{2}.$$

(b) Let $f(z) = z^2/(1+z^6)$. The function f is even, so

$$\int_{-\infty}^\infty f(x)dx = 2\int_0^\infty f(x)dx,$$

and we are reduced to computing the integral of f over the whole real line. Arguing like in (a) we see that we can use the same contour, hence

$$\int_{-\infty}^\infty f(x)dx = 2\pi i \sum \text{residues of } \frac{z^2}{1+z^6} \text{ in the upper half plane.}$$

The poles of f are described in part (a) of Exercise 1. Taking into account that z^2 is entire we can compute the residues at the poles and obtain

$$\int_{-\infty}^{\infty} f(x)dx = 2\pi i \left(\frac{e^{2\pi i/6}}{6e^{5\pi i/6}} + \frac{e^{2\pi i/2}}{6e^{5\pi i/2}} + \frac{e^{10\pi i/6}}{6e^{25\pi i/6}} \right)$$

$$= \frac{\pi i}{3}(e^{-\pi i/2} + e^{-3\pi i/2} + e^{-\pi i/2})$$

$$= \frac{\pi i}{3}(-i + i - i) = \frac{\pi}{3}.$$

The above observation implies that

$$\int_{0}^{\infty} f(x)dx = \frac{\pi}{6},$$

as was to be shown.

Exercise VI.2.3. *Show that*

$$\int_{-\infty}^{\infty} \frac{x-1}{x^5-1}dx = \frac{4\pi}{5}\sin\frac{2\pi}{5}.$$

Solution. Let $f(z) = (z-1)/(z^5-1)$. Then there exists a positive constant B such that for all large R we have

$$|f(z)| \leq B\frac{R^2}{R^5} = \frac{B}{R^3}$$

whenever $|z| = R$. The same argument as in Exercise 1 (a) shows that we can use the same contour as this exercise, therefore

$$\int_{-\infty}^{\infty} \frac{x-1}{x^5-1}dx = 2\pi i \sum \text{ residues of } \frac{z-1}{z^5-1} \text{ in the upper half plane.}$$

The simple poles of f in the upper half plane are at the points $e^{2\pi i/5}$ and $e^{4\pi i/5}$, so the residues at these points are

$$\frac{e^{2\pi i/5} - 1}{5(e^{2\pi i/5})^4} = \frac{e^{4\pi i/5} - e^{2\pi i/5}}{5} \quad \text{and} \quad \frac{e^{4\pi i/5} - 1}{5(e^{4\pi i/5})^4} = \frac{e^{8\pi i/5} - e^{4\pi i/5}}{5}.$$

Therefore

$$\int_{-\infty}^{\infty} \frac{x-1}{x^5-1}dx = \frac{2\pi i}{5}(e^{4\pi i/5} - e^{2\pi i/5} + e^{8\pi i/5} - e^{4\pi i/5})$$

$$= \frac{2\pi i}{5}(-e^{2\pi i/5} + e^{-2\pi i/5})$$

$$= \frac{2\pi i}{5} \cdot 2i \sin(2\pi/5)$$

$$= \frac{2\pi i}{5} \cdot -2i \sin(-2\pi/5)$$

$$= \frac{4\pi}{5}\sin(2\pi/5)$$

as was to be shown.

Exercise VI.2.4. *Evaluate*

$$\int_\gamma \frac{e^{-z^2}}{z^2} dz,$$

where γ *is:*
(a) the square with vertices $1+i, -1+i, -1-i, 1-i$.
(b) the ellipse defined by the equation

$$\frac{x^2}{a^2} + \frac{y^2}{b^2} = 1.$$

(The answer is 0 in both cases.)

Solution. The only singularity of the function e^{-z^2}/z^2 is at the origin. The power series expansion for the exponential gives

$$\frac{e^{-z^2}}{z^2} = \frac{1}{z^2} - 1 + \frac{z^2}{2!} - \frac{z^4}{3!}$$

so 0 is a pole of order 2. From the above expression we also see that the residue of e^{-z^2}/z^2 at the origin is 0. By the residue formula we conclude that the answer to (a) and (b) is 0.

Exercise VI.2.5. *(a)* $\int_{-\infty}^{\infty} \frac{e^{iax}}{x^2+1} dx = \pi e^{-a}$ *if* $a > 0$.
(b) For any real number $a > 0$,

$$\int_{-\infty}^{\infty} \frac{\cos x}{x^2 + a^2} dx = \pi e^{-a}/a.$$

[Hint: This is the real part of the integral obtained by replacing $\cos x$ *by* e^{ix}.]

Solution. (a) This integral belongs to the section on Fourier transforms: We must show that $f(z) = 1/(1+z^2)$ goes to 0 fast enough. There exists a constant K such that for all sufficiently large $|z|$ we have

$$|f(z)| \leq \frac{K}{|z|^2},$$

so the decay assumption is satisfied and we can use the formula given in the text (Theorem 2.2)

$$\int_{-\infty}^{\infty} \frac{e^{iax}}{1+x^2} dx = 2\pi i \sum \text{ residues of } e^{iaz} f(z) \text{ in the upper half plane.}$$

The function f has a simple pole at i with residue $1/2i$, so

$$\int_{-\infty}^{\infty} \frac{e^{iax}}{1+x^2} dx = 2\pi i \left(\frac{e^{iai}}{2i}\right) = \pi e^{-a},$$

as was to be shown.
(b) Changing variables $x = ay$ we get

$$\int_{-\infty}^{\infty} \frac{\cos x}{x^2 + a^2} dx = \frac{1}{a} \int_{-\infty}^{\infty} \frac{\cos(ay)}{y^2 + 1} dy$$

$$= \frac{1}{a}\operatorname{Re}\left(\int_{-\infty}^{\infty}\frac{e^{iay}}{y^2+1}dy\right)$$
$$= \frac{1}{a}\pi e^{-a},$$

as was to be shown.

Exercise VI.2.6. *Let* $a, b > 0$. *Let* $T \geq 2b$. *Show that*

$$\left|\frac{1}{2\pi i}\int_{-T}^{T}\frac{e^{iaz}}{z-ib}dz - e^{-ba}\right| \leq \frac{1}{Ta}(1 - e^{-Ta}) + e^{-Ta}.$$

Formulate a similar estimate when $a < 0$.

Solution. Let $f(z) = e^{iaz}/(z - ib)$. Consider the rectangle:

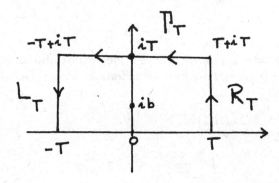

The only pole of f in this rectangle is at ib and the residue is e^{-ab}, so it suffices to show that

$$\frac{1}{2\pi i}\left|\int_{R_T}f + \int_{L_T}f + \int_{\Gamma_T}f\right| \leq \frac{1}{Ta}(1 - e^{-Ta}) + e^{-Ta},$$

where R_T denotes the right vertical segment, L_T the left vertical segment and Γ_T the top vertical segment (all with the orientation given on the picture). We begin with

$$\frac{1}{2\pi i}\int_{R_T}f = \frac{1}{2\pi i}\int_0^T\frac{e^{ia(T+it)}}{T+it-ib}i\,dt.$$

Putting absolute values we get

$$\left|\frac{1}{2\pi i}\int_{R_T}f\right| \leq \frac{1}{2\pi T}\int_0^T e^{-at}dt = \frac{1}{2\pi Ta}(1 - e^{-aT}).$$

The same estimate holds for the left hand side, namely

$$\left|\frac{1}{2\pi i}\int_{L_T}f\right| \leq \frac{1}{2\pi Ta}(1 - e^{-aT}).$$

We now estimate the integral on the top segment. With the parametrization $t + iT$, $-T \leq t \leq T$ we get

$$\left| \frac{1}{2\pi i} \int_{\Gamma_T} f \right| \leq \frac{e^{-aT}}{2\pi} \int_{-T}^{T} \frac{dt}{|t + iT - ib|}$$

$$\leq \frac{e^{-aT}}{2\pi} \frac{2T}{T - b}.$$

Since $T > 2b$, we must have $2T/(T - b) \leq 4$ so that

$$\left| \frac{1}{2\pi i} \int_{\Gamma_T} f \right| \leq \frac{2e^{-aT}}{\pi}.$$

We see now that our estimate is sharper than the one we wanted to prove.

If a is negative, then a similar argument with a rectangle lying in the lower half plane gives

$$\left| \frac{1}{2\pi i} \int_{-T}^{T} \frac{e^{iaz}}{z - ib} dz - e^{-ba} \right| \leq \frac{1}{Ta}(e^{aT} - 1) + e^{Ta}.$$

Exercise VI.2.7. *Let $c > 0$ and $a > 0$. Taking the integral over the vertical line, prove that*

$$\frac{1}{2\pi i} \int_{c-i\infty}^{c+i\infty} \frac{a^z}{z} dz = \begin{cases} 0 & \text{if } a < 1, \\ \frac{1}{2} & \text{if } a = 1, \\ 1 & \text{if } a > 1. \end{cases}$$

If $a = 1$, the integral is to be interpreted as the limit

$$\int_{c-i\infty}^{c+i\infty} = \lim_{T \to \infty} \int_{c-iT}^{c+iT}.$$

[Hint: If $a > 1$, integrate around a rectangle with corners $c - Ai$, $c + Bi$, $-X + Bi$, $-X - Ai$, and let $X \to \infty$. If $a < 1$, replace $-x$ by x.]

Solution. Let $b = \log a$ so that

$$f(z) = \frac{a^z}{z} = \frac{e^{bz}}{z}.$$

We begin with the case $a = 1$. Then $b = 0$ and we must evaluate the integral

$$\int_{c-i\infty}^{c+i\infty} \frac{1}{z} dz.$$

If $X > 0$, the segment from $c - iX$ to $c + iX$ is parametrized by $c + it$ where $-X \leq t \leq X$, so that

$$\int_{c-i\infty}^{c+i\infty} \frac{1}{z} dz = \int_{-X}^{X} \frac{i}{c + it} dt.$$

Now

$$\int_{-X}^{X} \frac{i}{c+it}\,dt = i \int_{-X}^{X} \frac{c}{c^2+t^2}\,dt + \int_{-X}^{X} \frac{t}{c^2+t^2}\,dt$$
$$= 2i \arctan(X/c)$$

so letting $X \to \infty$ we obtain

$$\int_{c-i\infty}^{c+i\infty} \frac{1}{z}\,dz = 2i\frac{\pi}{2} = i\pi$$

and this proves that

$$\frac{1}{2\pi i} \int_{c-i\infty}^{c+i\infty} \frac{a^z}{z}\,dz = \frac{1}{2}.$$

We now look at the case $a > 1$ or equivalently $b > 0$. Suppose $X > 0$ is large, and consider the contour:

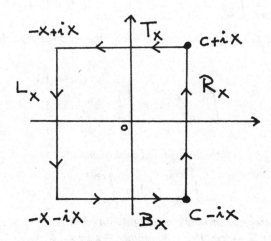

Here, T_X denotes the horizontal segment on top, B_X the horizontal segment on the bottom, L_X the vertical segment on the left and R_X the vertical segment on the right and all segments have the orientation given on the picture. If γ is the path defined by

$$\gamma = R_X + T_X + L_X + B_X$$

the residue formula gives

$$\frac{1}{2\pi i} \int_{\gamma} \frac{a^z}{z}\,dz = \sum \text{residues of } f \text{ in } \gamma.$$

The only pole of f is at the origin and since the numerator is equal to 1 at 0 we conclude that the right hand side of the above equality is equal to 1. Therefore, it suffices to show that the integral over T_X, L_X and B_X go to 0 as $X \to \infty$. We begin with T_X. This segment is parametrized by $t + iX$ with $-X \leq t \leq c$ so that

$$\int_{T_X} \frac{a^z}{z} dz = \int_c^{-X} \frac{e^{b(t+iX)}}{t+iX} dt,$$

and therefore

$$\left| \int_{T_X} \frac{a^z}{z} dz \right| \leq \int_{-X}^c \frac{e^{bt}}{|t+iX|} dt$$

$$\leq \frac{1}{X} \int_{-X}^c e^{bt} dt = \frac{1}{Xb} \left[e^{bc} - e^{-bX} \right],$$

which implies that

$$\left| \int_{T_X} \frac{a^z}{z} dz \right| \to 0 \quad \text{as } X \to \infty.$$

For L_X, we use the parametrization $-X + it$ where $-X \leq t \leq X$ so that

$$\int_{L_X} \frac{a^z}{z} dz = \int_{-X}^X i \frac{e^{b(-X+it)}}{-X+it} dt.$$

Therefore

$$\left| \int_{L_X} \frac{a^z}{z} dz \right| \leq \int_{-X}^X \frac{e^{-bX}}{|t+iX|} dt$$

$$\leq \frac{e^{-bX}}{X} \int_{-X}^X dt \leq 2e^{-bX},$$

and this proves that

$$\left| \int_{L_X} \frac{a^z}{z} dz \right| \to 0 \quad \text{as } X \to \infty.$$

Finally, we must show that the integral over B_X tends to 0 as $X \to \infty$. To do this, we use the parametrization $t - iX$ where $-X \leq t \leq c$, and estimating as before one easily finds that

$$\left| \int_{B_X} \frac{a^z}{z} dz \right| \leq \frac{1}{Xb} \left[e^{bc} - e^{-bX} \right],$$

and this settles the case $a > 1$.

For the case $a < 1$ or equivalently $b < 0$ we consider the following contour:

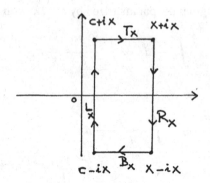

If $\gamma = R_X + T_X + L_X + B_X$, then the residue formula gives

$$\frac{1}{2\pi i}\int_\gamma \frac{a^z}{z}dz = \sum \text{residues of } f \text{ in } \gamma.$$

so it suffices to show that the integral over R_X, T_X and B_X tend to 0 as $X \to \infty$. To prove this, we argue as before. With the obvious parametrizations we obtain

$$\left|\int_{T_X}\frac{a^z}{z}dz\right| \le \frac{1}{bX}\left[e^{bX} - e^{bc}\right],$$

and the right hand side goes to 0 as $X \to \infty$. Similarly, we obtain that

$$\left|\int_{B_X}\frac{a^z}{z}dz\right| \to 0 \quad \text{and} \quad \left|\int_{R_X}\frac{a^z}{z}dz\right| \to 0$$

as $X \to 0$ and this concludes the proof.

Exercise VI.2.8. *(a) Show that for $a > 0$ we have*

$$\int_{-\infty}^{\infty}\frac{\cos x}{(x^2 + a^2)^2}dx = \frac{\pi(1 + a)}{2a^3 e^a}.$$

(b) Show that for $a > b > 0$ we have

$$\int_0^{\infty}\frac{\cos x}{(x^2 + a^2)(x^2 + b^2)}dx = \frac{\pi}{a^2 - b^2}\left(\frac{1}{be^b} - \frac{1}{ae^a}\right).$$

Solution. The function $\sin x$ is odd so $\int_{-\infty}^{\infty}\sin x/(x^2 + a^2)^2 dx = 0$ and therefore

$$\int_{-\infty}^{\infty}\frac{\cos x}{(x^2 + a^2)^2}dx = \int_{-\infty}^{\infty}\frac{e^{ix}}{(x^2 + a^2)^2}dx.$$

Let $f(z) = 1/(z^2 + a^2)^2$. We want to find the Fourier transform $\int_{-\infty}^{\infty}f(x)e^{ix}dx$. An estimate like in Exercise 5 shows that we can apply Theorem 2.2, and therefore

$$\int_{-\infty}^{\infty}f(x)e^{ix}dx = 2\pi i \sum \text{residues of } f(z)e^{iz} \text{ in the upper half plane.}$$

The only pole of f in the upper half plane is at ia. We must now find the residue of f at this pole. We write

$$f(z) = \frac{1}{(z - ia)^2(z + ia)^2}.$$

Now we have

$$(z + ia)^{-2} = (z - ia + 2ia)^{-2} = (2ia)^{-2}\left(1 + \frac{z - ia}{2ia}\right)^{-2}$$

which after expanding becomes

$$(z + ia)^{-2} = (2ia)^{-2}\left(1 - 2\frac{z - ia}{2ia} + \cdots\right).$$

We also have $e^{iz} = e^{-a}e^{i(z-ia)} = e^{-a}(1 + i(z - ia) + \cdots)$ so

$$f(z)e^{iz} = \frac{e^{-a}}{(z - ia)^2(2ia)^2}\left(1 - 2\frac{z - ia}{2ia} + \cdots\right)(1 + i(z - ia) + \cdots).$$

Hence

$$\text{res}_{ia}\, f(z)e^{iz} = \frac{e^{-a}}{(2ia)^2}\left(\frac{-1}{ia} + i\right) = \frac{e^{-a}(1 + a)}{4a^3 i}.$$

By the residue formula we conclude that

$$\int_{-\infty}^{\infty} f(x)e^{ix}dx = 2\pi i\,\frac{e^{-a}(1 + a)}{4a^3 i} = \pi\,\frac{e^{-a}(1 + a)}{2a^3}$$

as was to be shown.

(b) Arguing like in (a) and using the fact that cos is even we find that the desired integral is equal to $\frac{1}{2}\int_{-\infty}^{\infty} f(x)e^{ix}dx$ where

$$f(z) = \frac{1}{(z^2 + a^2)(z^2 + b^2)}.$$

We can apply Theorem 2.2. We are only concerned with singularities in the upper half plane. In this region f has two simple poles one at ia and the other at ib. Computing the derivative of $(z^2 + a^2)$ implies that the residue of $f(z)e^{iz}$ at ia is

$$\text{res}_{z=ia}\, f(z)e^{iz} = \frac{e^{i(ia)}}{(2ia)((ia)^2 + b^2)} = -\frac{e^{-a}}{(2ia)(a^2 - b^2)}.$$

Similarly we find that

$$\text{res}_{z=ib}\, f(z)e^{iz} = \frac{e^{i(ib)}}{(a^2 + (ib)^2)(2ib)} = -\frac{e^{-b}}{(2ib)(a^2 - b^2)}.$$

By Theorem 2.2 we obtain

$$\int_{-\infty}^{\infty} f(x)e^{ix}dx = 2\pi i(\text{res}_{z=ia}\, f(z)e^{iz} + \text{res}_{z=ib}\, f(z)e^{iz})$$

$$= \frac{\pi}{a^2 - b^2}\left(-\frac{e^{-a}}{a} + \frac{e^{-b}}{b}\right).$$

Conclude.

Exercise VI.2.9. $\int_0^\infty \frac{\sin^2 x}{x^2} dx = \pi/2$. *[Hint: Consider the integral of* $(1 - e^{2ix})/x^2$.*]*

Solution. Since the integrand is even, the desired integral is equal to

$$\frac{1}{2} \int_{-\infty}^\infty \frac{\sin^2 x}{x^2} dx.$$

The trigonometric identity $2 \sin^2 x = 1 - \cos 2x$, implies

$$2 \int_{-\infty}^\infty \frac{\sin^2 x}{x^2} dx = \text{Re} \left(\int_{-\infty}^\infty \frac{1 - e^{2ix}}{x^2} dx \right).$$

We have reduced the problem to finding the integral $\int_{-\infty}^\infty f(x)dx$ where $f(z) = (1 - e^{2iz})/z^2$. The function f has a unique pole at the origin. We take as a path

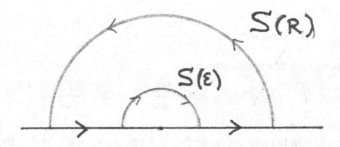

To show that

$$\lim_{R \to \infty} \int_{S(R)} f(z)dz = 0$$

split the integral and write is as

$$\int_{S(R)} \frac{dz}{z^2} - \int_{S(R)} \frac{e^{2iz}}{z^2} dz.$$

The first integral goes to 0 as R tends to infinity because it is bounded by $\pi R/R^2$, namely the sup norm of $1/z^2$ on $S(R)$ times the length of $S(R)$. The second integral is estimated exactly like on page 196 of Lang's book. By the lemma on this same page we obtain

$$\lim_{\epsilon \to 0} \int_{S(\epsilon)} f(z)dz = -\pi i \, \text{res}_{z=0} f(z).$$

To find the residue, we must use the power series expansion of the exponential

$$f(z) = \frac{1 - (1 + 2iz + (2iz)^2/2! + \cdots)}{z^2} = \frac{-2i}{z} + \text{terms of higher order.}$$

Hence the residue of f at the origin is $-2i$ and therefore

$$\int_{-\infty}^{\infty} \frac{1-e^{2ix}}{x^2}dx = 2\pi.$$

Conclude.

Exercise VI.2.10. $\int_{-\infty}^{\infty} \frac{\cos x}{a^2-x^2}dx = \frac{\pi \sin a}{a}$ *for* $a > 0$. *The integral is meant to be interpreted as the limit:*

$$\lim_{B\to\infty} \lim_{\delta\to 0} \int_{-B}^{-a-\delta} + \int_{-a+\delta}^{a-\delta} + \int_{a+\delta}^{B}.$$

Solution. Since the sine function is odd, the integral we must compute is equal to

$$\int_{-\infty}^{\infty} f(x)dx \quad \text{where } f(z) = \frac{e^{iz}}{a^2-z^2}.$$

The function f has two simple poles, one at a and the other at $-a$. Consider the following contour:

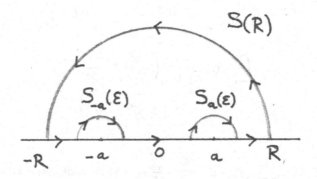

We must show that

$$\lim_{R\to\infty} \int_{S(R)} f(z)dz = 0.$$

We argue like on page 196 of Lang's book. We have

$$\int_{S(R)} f(z)dz = \int_0^\pi \frac{e^{iR\cos\theta}e^{-R\sin\theta}}{a^2 - R^2e^{2i\theta}} iRe^{i\theta}d\theta,$$

so for all large R we get

$$\left|\int_{S(R)} f(z)dz\right| \le \int_0^\pi \frac{e^{-R\sin\theta}}{R^2-a^2}Rd\theta = \frac{2R}{R^2-a^2}\int_0^{\pi/2} e^{-R\sin\theta}d\theta.$$

But if $0 \leq \theta \leq \pi/2$, then $\sin \theta \geq 2\theta/\pi$, thus

$$\left| \int_{S(R)} f(z)dz \right| \leq \frac{2R}{R^2 - a^2} \int_0^{\pi/2} e^{-2R\theta/\pi} d\theta = \frac{\pi}{R^2 - a^2}(1 - e^{-R}),$$

and now it is clear that our limit holds.

Now we must evaluate the limits

$$\lim_{\epsilon \to 0} \int_{S_a(\epsilon)} f(z)dz \quad \text{and} \quad \lim_{\epsilon \to 0} \int_{S_{-a}(\epsilon)} f(z)dz.$$

A simple modification of the lemma on page 196 of Lang's book shows that if f has a pole at x, then

$$\lim_{\epsilon \to 0} \int_{S_x(\epsilon)} f(z)dz = \pi \operatorname{res}_{z=x} f(z).$$

Writing f as

$$f(z) = \frac{e^{iz}}{(a - z)(a + z)}$$

we find that

$$\operatorname{res}_{z=a} f(z) = \frac{-e^{ia}}{2a} \quad \text{and} \quad \operatorname{res}_{z=-a} f(z) = \frac{e^{-ia}}{2a}.$$

Therefore

$$\int_{-\infty}^{\infty} f(z)dz = \pi \left(\frac{-e^{ia}}{2a} + \frac{e^{-ia}}{2a} \right)$$

$$= \frac{\pi}{a} \left(\frac{e^{ia}}{2i} - \frac{e^{-ia}}{2i} \right)$$

$$= \frac{\pi \sin a}{a}.$$

Exercise VI.2.11. $\int_{-\infty}^{\infty} \frac{\cos x}{e^x + e^{-x}} dx = \frac{\pi}{e^{\pi/2} + e^{-\pi/2}}$. *Use the indicated contour:*

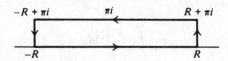

Solution. The sine function is odd, so the desired integral is equal to

$$\int_{-\infty}^{\infty} f(x)dx \quad \text{where } f(z) = \frac{e^{iz}}{e^z + e^{-z}}.$$

To find the singularities of f we must solve $e^z + e^{-z} = 0$. Multiplying this equation by e^z we get $e^{2z} + 1 = 0$. Letting $z = x + iy$, we get $e^{2x} e^{2iy} = -1$.

Putting absolute values we find $x = 0$ and this shows that f has singularities at the points $i(\pi/2 + k\pi)$ where $k \in \mathbf{Z}$.

Consider the contour $\gamma(R) = \gamma_1(R) + \gamma_2(R) + \gamma_3(R) + \gamma_4(R)$ as shown on the figure

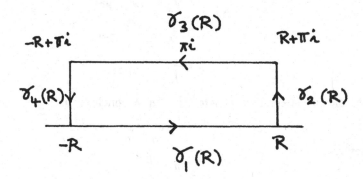

The only singularity of f in the interior of the contour is at $i\pi/2$. The derivative of $e^z + e^{-z}$ at that point is equal to $2i$ which is nonzero so f has a simple pole at $i\pi/2$ with

$$\operatorname{res}_{z=i\pi/2} f(z) = \frac{e^{i(i\pi/2)}}{2i} = \frac{e^{-\pi/2}}{2i}.$$

By the residue formula, we get

$$\int_{\gamma(R)} f(z)dz = \pi e^{-\pi/2}.$$

We now want show that the integral over $\gamma_2(R)$ and $\gamma_4(R)$ tend to 0 as R tends to infinity. We can estimate the integral by

$$\left| \int_{\gamma_2(R)} f(z)dz \right| \leq \int_{\gamma_2(R)} |f(z)| d \leq \pi \sup_{0 \leq y \leq \pi} \left| \frac{e^{iR}e^{-y}}{e^R e^{iy} + e^{-R}e^{-iy}} \right|,$$

and for large R

$$\left| \frac{e^{iR}e^{-y}}{e^R e^{iy} + e^{-R}e^{-iy}} \right| \leq \frac{e^{-y}}{e^R \left| e^{iy} + e^{-2R}e^{-iy} \right|} \leq \frac{1}{e^R(1 - e^{-2R})}.$$

The last inequality follows from $0 \leq y \leq \pi$ and the triangle inequality applied to the denominator and the fact that R is large. It is now clear that the integral of f over $\gamma_2(R)$ tends to 0 as R tends to infinity. A similar argument proves the same result for the integral of f over $\gamma_4(R)$.

Finally, we find the expression of the integral of f over $\gamma_3(R)$. Using the parametrization $t + \pi$ for $-R \leq t \leq R$ and being careful about the orientation we get

$$\int_{\gamma_3(R)} f(z)dz = \int_R^{-R} \frac{e^{it+\pi}}{e^{t+\pi i} + e^{-t\pi i}} dt$$

$$= e^{-\pi} \int_R^{-R} \frac{e^{it}}{-e^t - e^{-t}} dt$$

$$= e^{-\pi} \int_{-R}^{R} \frac{e^{it}}{e^t + e^{-t}} dt$$

$$= e^{-\pi} \int_{\gamma_1(R)} f(z)dz.$$

So if I denotes the integral we want to evaluate we conclude that

$$I + e^{-\pi} I = \pi e^{-\pi/2},$$

and therefore

$$I = \frac{\pi}{e^{\pi/2} + e^{-\pi/2}}.$$

This concludes the exercise.

Exercise VI.2.12. $\int_0^\infty \frac{x \sin x}{x^2 + a^2} dx = \frac{1}{2}\pi e^{-a}$ if $a > 0$.

Solution. The integral we wish to evaluate has an even integrand so it is equal to

$$\frac{1}{2} \int_{-\infty}^{\infty} \frac{x \sin x}{x^2 + a^2} dx.$$

The function $x \cos x$ is odd so

$$\int_{-\infty}^{\infty} \frac{x \sin x}{x^2 + a^2} dx = \text{Im} \left(\int_{-\infty}^{\infty} f(x)e^{ix} dx \right) \quad \text{where } f(z) = \frac{z}{z^2 + a^2}.$$

Clearly, the function f verifies the hypothesis of Theorem 2.2 so we can apply the formula

$$\int_{-\infty}^{\infty} f(x)e^{ix} dx = 2\pi i \sum \text{residues of } f(z)e^{iz} \text{ in the upper half plane.}$$

The function f has simple poles at ia and $-ia$. Since $a > 0$ we are only concerned with the pole at ia which is in the upper half plane. Since

$$f(z) = \frac{z}{(z - ia)(z + ia)},$$

it follows that

$$\text{res}_{z=ia} f(z)e^{iz} = \left(\frac{ia}{2ia} \right) e^{i(ia)} = \frac{e^{-a}}{2}.$$

Hence

$$\int_{-\infty}^{\infty} f(x)e^{ix} dx = \pi i e^{-a}.$$

The observations at the beginning of the exercise imply that

$$\int_0^\infty \frac{x \sin x}{x^2 + a^2} dx = \frac{1}{2}\pi e^{-a}.$$

Exercise VI.2.13. $\int_{-\infty}^\infty \frac{e^{ax}}{e^x + 1} dx = \frac{\pi}{\sin \pi a}$ *for* $0 < a < 1$.

Solution. The solution to this exercise is very much like our answer to Exercise VI.2.11. Let $f(z) = e^{az}/(e^z + 1)$. The function f has poles at $i\pi + 2k\pi$ with $k \in \mathbf{Z}$. Consider the contour $\gamma(R) = \gamma_1(R) + \gamma_2(R) + \gamma_3(R) + \gamma_4(R)$ given by

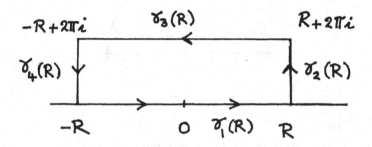

Taking the derivative of the denominator of f we find that the residue of f at $i\pi$ is $e^{ai\pi}/e^{i\pi} = -e^{ai\pi}$ so by the residue formula we obtain

$$\int_{\gamma(R)} f(z)dz = -2\pi i e^{ai\pi}.$$

We must show that the integrals on the sides $\gamma_2(R)$ and $\gamma_4(R)$ tend to 0 as R tends to infinity. We estimate the sup norm of f on $\gamma_2(R)$ by

$$\sup_{z \in \gamma_2(R)} |f(z)| = \sup_{z \in \gamma_2(R)} \left| \frac{e^{aR}e^{iay}}{e^R e^{iy} + 1} \right| \le \frac{e^{aR}}{e^R - 1}.$$

But $0 < a < 1$ so we see that the sup norm of f on $\gamma_2(R)$ goes to 0 as R tends to infinity, and since $\gamma_2(R)$ has length 2π we conclude that the integral of f over $\gamma_2(R)$ tends to 0 as R tends to infinity. A similar argument shows that the same conclusion holds for the integral of f over $\gamma_4(R)$.

We must now find an expression for the integral of f over $\gamma_3(R)$. Arguing like in Exercise 11 we find that

$$\int_{\gamma_3(R)} f(z)dz = -e^{2\pi a i} \int_{\gamma_1(R)} f(z)dz.$$

If I denotes the integral we want to compute, we get (letting $R \to \infty$)

$$I - e^{2\pi a i} I = -2\pi i e^{ai\pi}$$

so that

$$\frac{(e^{\pi ai} - e^{-\pi ai})}{2i} I = \pi.$$

We have therefore proved that $I = \pi/(\sin \pi a)$.

Exercise VI.2.14. (a) $\int_0^\infty \frac{(\log x)^2}{1+x^2} dx = \pi^3/8$. *Use the contour*

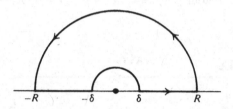

(b) $\int_0^\infty \frac{\log x}{(x^2+1)^2} dx = -\pi/4$.

Solution. (a) We first define the following mysterious function:

$$f(z) = \frac{(\log z - \frac{i\pi}{2})^2}{1 + z^2}.$$

We take the branch of the logarithm given by deleting the negative imaginary axis and taking the angle to be $-\pi/2 < \theta < 3\pi/2$. Consider the contour given by

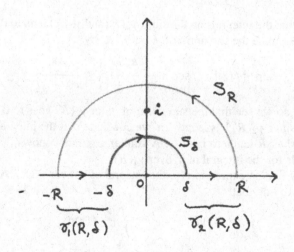

The only singularity of f which is of interest is the simple pole at i. The residue of f at that pole is

$$\frac{(\log i - i\pi/2)^2}{2i} = 0.$$

This is one reason which explains the strange constant $\pi i/2$ in the definition of f. By the residue formula, we conclude that $\int_\gamma f(z)dz = 0$. The integral of f on S_R tends to 0 as $R \to \infty$ because the length of S_R multiplied by the sup norm on S_R behaves like $R\frac{(\log R)^2}{R^2}$ which tends to 0 as R tends to infinity. The integral of f on S_δ behaves like $(\log \delta)^2 \delta$ which tends to 0 as $\delta \to 0$.

On the real axis we have

$$\int_{\gamma_1(R,\delta)} f(x)dx = \int_{-R}^{-\delta} \frac{(\log |x| + i(\pi/2))^2}{1+x^2} dx$$

and

$$\int_{\gamma_2(R,\delta)} f(x)dx = \int_\delta^R \frac{(\log |x| - i(\pi/2))^2}{1+x^2} dx.$$

Letting $R \to \infty$ and $\delta \to 0$ we see that after cancellations (which explain the choice of our f) we get

$$\int_{-\infty}^0 \frac{(\log |x|)^2}{1+x^2}dx + \int_0^\infty \frac{(\log |x|)^2}{1+x^2}dx - \frac{\pi^2}{4}\int_{-\infty}^\infty \frac{dx}{1+x^2} = 0,$$

hence

$$2\int_0^\infty \frac{(\log x)^2}{1+x^2}dx = \frac{\pi^2}{4}\int_{-\infty}^\infty \frac{dx}{1+x^2} = \frac{\pi^3}{4}.$$

(b) We use the same technique as in (a). Let

$$f(z) = \frac{\log z - \frac{i\pi}{2}}{(z^2+1)^2}.$$

We use the same branch of the logarithm and the same contour as in part (a). The only singularity of f in the upper half plane is at the point i. Our next step is to find the residue of f at this singularity. Since we can write

$$f(z) = \frac{\log z - \frac{i\pi}{2}}{(z+i)^2(z-i)^2}$$

it suffices to find the coefficient of the term $z - i$ in the power series expansion of $(\log z - i\pi/2)/(z+i)^2$ near i. We simply have

$$\frac{1}{(z+i)^2} = \frac{1}{(2i)^2\left(1 + \frac{z-i}{2i}\right)^2} = \frac{-1}{4}\left(1 - 2\frac{z-i}{2i} + \text{higher order terms}\right),$$

and

$$\log z - i\pi/2 = \sum \frac{(-1)^{n-1}}{n}\left(\frac{z-i}{i}\right)^n = \frac{z-i}{i} + \text{higher order terms}.$$

Thus

$$\operatorname{res}_{z=i} f(z) = \frac{-1}{4i}.$$

The residue formula gives

$$\int_{\gamma} f(z)dz = 2\pi i \operatorname{res}_{z=i} f(z) = \frac{-\pi}{2}.$$

An argument similar to the one given in (a) shows that the integrals on the semicircles S_R and S_δ tend to 0 as $R \to \infty$ and $\delta \to 0$ respectively. Therefore

$$\int_{-\infty}^{0} \frac{\log|x| + i\pi/2}{(x^2+1)^2} dx + \int_{0}^{\infty} \frac{\log|x| - i\pi/2}{(x^2+1)^2} dx = \frac{-\pi}{2}.$$

We obtain

$$2 \int_{0}^{\infty} \frac{\log x}{(x^2+1)^2} dx = \frac{-\pi}{2},$$

as was to be shown.

Exercise VI.2.15. (a) $\int_0^\infty \frac{x^a}{1+x} \frac{dx}{x} = \frac{\pi}{\sin \pi a}$ for $0 < a < 1$.
(b) $\int_0^\infty \frac{x^a}{1+x^3} \frac{dx}{x} = \frac{\pi}{3 \sin(\pi a/3)}$ for $0 < a < 3$.

Solution. Let $f(z) = 1/(1+z)$. Then $|f(z)| \le C/|z|$ as $|z| \to \infty$ for some constant C and $|f(z)| \to 1$ as $|z| \to 0$, so we can apply Theorem 2.4 which states that the integral (a Mellin transform)

$$\int_0^\infty f(x) x^a \frac{dx}{x}$$

is equal to $-\frac{\pi e^{-\pi i a}}{\sin \pi a}$ times the sum of the residues of $f(z)z^{a-1}$ at the poles of f, excluding the residue at 0.

The only pole of f is at -1 and

$$\operatorname{res}_{z=-1} f(z) z^{a-1} = (-1)^{a-1} = e^{(a-1)\log(-1)} = e^{(a-1)i\pi}.$$

Therefore

$$\int_0^\infty \frac{x^a}{1+x} \frac{dx}{x} = -\frac{\pi e^{-\pi i a}}{\sin \pi a} e^{(a-1)i\pi} = \frac{\pi}{\sin \pi a}.$$

(b) As in part (a), we can apply Theorem 2.4, so all we have to do is compute the residues of $f(z)z^{a-1}$ where $f(z) = 1/(1+z^3)$. The poles of f are at $e^{i\pi/3}$, $e^{i\pi}$ and $e^{5i\pi/3}$ so the sum of the residues of $f(z)z^{a-1}$ excluding the residue at the origin is

$$\frac{(e^{i\pi/3})^{a-1}}{3(e^{i\pi/3})^2} + \frac{(e^{i\pi})^{a-1}}{3(e^{i\pi})^2} + \frac{(e^{5i\pi/3})^{a-1}}{3(e^{5i\pi/3})^2}.$$

We transform the first term in the following way

$$\frac{(e^{i\pi/3})^{a-1}}{3(e^{i\pi/3})^2} = e^{(a-1)(i\pi/3)} 3 e^{2i\pi/3} = \frac{e^{ai\pi/3}e^{-i\pi}}{3} = -\frac{e^{ai\pi/3}}{3}.$$

Making the same transformations to the other terms, we find that the sum of the residues of $f(z)z^{a-1}$ excluding the residue at the origin is

$$= \frac{-1}{3} \left(e^{ai\pi/3} + e^{ai\pi} + e^{ai5\pi/3} \right)$$

$$= \frac{-e^{ai\pi}}{3} \left(e^{ai(-2)\pi/3} + 1 + e^{ai2\pi/3} \right).$$

Hence

$$\int_0^\infty \frac{x^a}{1+x^3} \frac{dx}{x} = \frac{\pi}{3 \sin \pi a} \left(e^{ai(-2)\pi/3} + 1 + e^{ai2\pi/3} \right).$$

We claim that

$$\frac{e^{ai(-2)\pi/3} + 1 + e^{ai2\pi/3}}{\sin \pi a} = \frac{1}{\sin(\pi a/3)}.$$

Using Euler's formula $2i \sin \theta = e^{i\theta} - e^{-i\theta}$ to write everything with exponentials and cross multiplying proves our claim.

Exercise VI.2.16. *Let f be a continuous function, and suppose that the integral*

$$\int_0^\infty f(x)x^a \frac{dx}{x}$$

is absolutely convergent. Show that it is equal to the integral

$$\int_{-\infty}^\infty f(e^t)e^{at} dt.$$

If we put $g(t) = f(e^t)$, this shows that the Mellin transform is essentially a Fourier transform, up to a change of variable.

Solution. We change variables $e^t = x$. Then $dx = e^t dt$ and therefore

$$\int_0^\infty f(x)x^a \frac{dx}{x} = \int_{-\infty}^\infty f(e^t)(e^t)^a e^t \frac{dt}{e^t} = \int_{-\infty}^\infty f(e^t)e^{at} dt.$$

Exercise VI.2.17. $\int_0^{2\pi} \frac{1}{1+a^2-2a\cos\theta} d\theta = \frac{2\pi}{1-a^2}$ *if $0 < a < 1$. The answer comes out to the negative of that if $a > 1$.*

Solution. Since this is a trigonometric integral we will apply Theorem 2.3. We have

$$f(z) = \frac{1}{iz} \frac{1}{1 + a^2 - 2a \left(\frac{1}{2}(z + \frac{1}{z}) \right)} = \frac{1}{i} \frac{1}{-az^2 + (1+a^2)z - a}.$$

The roots of the denominator of the second fraction are

$$z_1 = \frac{-(1+a^2) + \sqrt{(1-a^2)^2}}{-2a} \quad \text{and} \quad z_2 = \frac{-(1+a^2) - \sqrt{(1-a^2)^2}}{-2a}.$$

If $0 < a < 1$, the only pole of f in the unit circle is at $z_1 = a$ and (differentiating the denominator of the fraction) we find that the residue is

$$\frac{1}{i}\frac{1}{-2az^1 + (1 + a^2)} = \frac{1}{i(1 - a^2)},$$

and therefore

$$\int_C f(z)dz = 2\pi i \left(\frac{1}{i(1 - a^2)}\right) = \frac{2\pi}{1 - a^2}.$$

If $a > 1$ the only pole of f in the unit circle is at $z_1 = 1/a$ and the residue is

$$\frac{1}{i}\frac{1}{-2az^1 + (1 + a^2)} = \frac{1}{i(-1 + a^2)},$$

hence

$$\int_C f(z)dz = \frac{2\pi}{a^2 - 1}.$$

Exercise VI.2.18. $\int_0^\pi \frac{1}{1+\sin^2\theta}d\theta = \frac{\pi}{\sqrt{2}}.$

Solution. See Exercise 20.

Exercise VI.2.19. $\int_0^\pi \frac{1}{3+2\cos\theta}d\theta = \frac{\pi}{\sqrt{5}}.$

Solution. In order to apply Theorem 2.3 we must integrate from 0 to 2π. We claim that

$$\int_0^\pi \frac{1}{3 + 2\cos\theta}d\theta = \frac{1}{2}\int_0^{2\pi} \frac{1}{3 + 2\cos\theta}d\theta.$$

To prove this claim, we change variables $\theta \to -\theta$ in the first integral so that

$$\int_0^\pi \frac{1}{3 + 2\cos\theta}d\theta = \int_0^{-\pi} \frac{-1}{3 + 2\cos(-\theta)}d\theta = \int_{-\pi}^0 \frac{1}{3 + 2\cos\theta}d\theta.$$

Now changing variables $\theta \to \theta + 2\pi$ we get

$$\int_{-\pi}^0 \frac{1}{3 + 2\cos\theta}d\theta = \int_\pi^{2\pi} \frac{1}{3 + 2\cos\theta}d\theta.$$

This proves our claim. We must now compute

$$\int_0^{2\pi} \frac{1}{3 + 2\cos\theta}d\theta$$

and we use Theorem 2.3 with the function

$$f(z) = \frac{1}{iz}\frac{1}{3 + 2\frac{1}{2}\left(z + \frac{1}{z}\right)} = \frac{1}{i(z^2 + 3z + 1)}.$$

The zeros of the denominator are

$$z_1 = \frac{-3 + \sqrt{5}}{2} \quad \text{and} \quad z_2 = \frac{-3 - \sqrt{5}}{2}.$$

The only pole of f in the unit circle is at z_1 and the residue is

$$\frac{1}{i(2z_1 + 3)} = \frac{1}{i\sqrt{5}},$$

and therefore

$$\int_0^{2\pi} \frac{d\theta}{3 + 2\cos\theta} = 2\pi i \frac{1}{2i\sqrt{5}} = \frac{2\pi}{\sqrt{5}}.$$

This proves that

$$\int_0^{\pi} \frac{d\theta}{3 + 2\cos\theta} = \frac{\pi}{\sqrt{5}}.$$

Exercise VI.2.20. $\int_0^{\pi} \frac{a\,d\theta}{a^2 + \sin^2\theta} = \int_0^{2\pi} \frac{a\,d\theta}{1 + 2a^2 - \cos\theta} = \frac{\pi}{\sqrt{1+a^2}}.$

Solution. We have

$$a^2 + \sin^2\theta = a^2 + \frac{1 - \cos 2\theta}{2} = \frac{1}{2}\left(2a^2 + 1 - \cos 2\theta\right),$$

so changing variables $\varphi = 2\theta$ we find that

$$\int_0^{\pi} \frac{a\,d\theta}{a^2 + \sin^2\theta} = \int_0^{2\pi} \frac{a\,d\theta}{1 + 2a^2 - \cos\theta} = \frac{\pi}{\sqrt{1+a^2}}.$$

To compute this last integral, we use Theorem 2.3 with

$$f(z) = \frac{1}{iz} \frac{a}{1 + 2a^2 - \left(\frac{1}{2}\left(z + \frac{1}{z}\right)\right)} = \frac{2ai}{z^2 - (2 + 4a^2)z + 1}.$$

The roots of the denominator are

$$z_1 = \frac{2 + 4a^2 + \sqrt{16a^2 + 16a^4}}{2} = 1 + 2a^2 + 2|a|\sqrt{1 + a^2},$$

and

$$z_2 = 1 + 2a^2 - 2|a|\sqrt{1 + a^2}.$$

The only pole of f in the unit circle is at z_2 and the residue of f at this point is

$$\frac{2ai}{2z_2 - (2 + 4a^2)} = \frac{ai}{-2|a|\sqrt{1 + a^2}}$$

and therefore

$$\int_C f(z)\,dz = 2\pi i \frac{ai}{-2|a|\sqrt{1 + a^2}} = \frac{a}{|a|} \frac{\pi}{\sqrt{1 + a^2}}.$$

Conclude.

Exercise VI.2.21. $\int_0^{\pi/2} \frac{1}{(a + \sin^2\theta)^2} d\theta = \frac{\pi(2a+1)}{4(a^2 + a)^{3/2}}$ *for a > 0.*

Solution. Using the fact that

$$\sin^2\theta = \frac{1}{2}(1 - \cos 2\theta)$$

and arguing like at the beginning of Exercise 19, one finds after a few linear changes of variables that

$$\int_0^{\pi/2} \frac{1}{(a + \sin^2 \theta)^2} d\theta = \int_0^{2\pi} \frac{d\theta}{(2a + 1 - \cos \theta)^2}.$$

Since we reduced the problem to a trigonometric integral from 0 to 2π we can apply Theorem 2.3 with the function.

$$f(z) = \frac{1}{iz} \frac{1}{\left(2a + 1 - \frac{1}{2}\left(z + \frac{1}{z}\right)\right)^2}$$

$$= \frac{z}{i\left(-\frac{z^2}{2} + (2a + 1)z - \frac{1}{2}\right)^2}.$$

The zeros of the denominator are at the points

$$z_1 = (2a + 1) - 2\sqrt{a^2 + a} \quad \text{and} \quad z_2 = (2a + 1) + 2\sqrt{a^2 + a}.$$

Since z_1 is the only pole of f in the unit circle we must compute the residue of f at this point. We write

$$f(z) = \frac{z}{i(1/4)(z - z_1)^2(z - z_2)^2} = \frac{4z}{i(z - z_1)^2(z - z_2)^2},$$

so that the residue of f is equal to the coefficient of $z - z_1$ in the power series expansion of

$$h(z) = \frac{4z}{i(z - z_2)^2}$$

near z_1. To find this coefficient, we first differentiate h and obtain

$$h'(z) = \frac{4}{i}\left[\frac{1}{(z - z_2)^2} - 2\frac{z}{(z - z_2)^3}\right] = \frac{4}{i}\left[\frac{-z - z^2}{(z - z_2)^3}\right],$$

which we evaluate at z_1 to obtain the residue of f at z_1

$$\text{res}_{z=z_1} f(z) = h'(z_1) = \frac{4}{i}\frac{-4a - 2}{-4^3(a^2 + a)^{3/2}} = \frac{1}{8i}\frac{2a + 1}{(a^2 + a)^{3/2}}.$$

Therefore

$$\int_C f(z)dz = 2\pi i \frac{1}{8i}\frac{2a + 1}{(a^2 + a)^{3/2}} = \frac{\pi(2a + 1)}{4(a^2 + a)^{3/2}}.$$

Exercise VI.2.22. $\int_0^{2\pi} \frac{1}{2 - \sin \theta} d\theta = 2\pi/\sqrt{3}$.

Solution. We will apply Theorem 2.3 with the function

$$f(z) = \frac{1}{iz}\frac{1}{2 - \frac{1}{2i}\left(z - \frac{1}{z}\right)} = \frac{2}{-z^2 + 4iz + 1}.$$

The roots of the denominator are

$$z_1 = 2i - i\sqrt{3} \quad \text{and} \quad z_2 = 2i + i\sqrt{3}.$$

The only pole of f in the unit circle is at z_1 and the residue of f at this point is

$$\frac{2}{-2z_1 + 4i} = \frac{1}{i\sqrt{3}}.$$

Hence

$$\int_C f(z)dz = 2\pi i \frac{1}{i\sqrt{3}} = \frac{2\pi}{\sqrt{3}}.$$

Exercise VI.2.23. $\int_0^{2\pi} \frac{1}{(a+b\cos\theta)^2}d\theta = \frac{2\pi a}{(a^2-b^2)^{3/2}}$ for $0 < b < a$.

Solution. We will apply Theorem 2.3 with

$$f(z) = \frac{1}{iz}\frac{1}{\left(a + \frac{b}{2}\left(z + \frac{1}{z}\right)\right)^2} = \frac{z}{i\left(\frac{b}{2}z^2 + az + \frac{b}{2}\right)^2}.$$

The roots of the denominator are

$$z_1 = \frac{-a + \sqrt{a^2 - b^2}}{b} \quad \text{and} \quad z_2 = \frac{-a - \sqrt{a^2 - b^2}}{b}.$$

The assumption that $0 < b < a$ implies that the only pole of f in the unit circle is at z_1. We must now compute the residue of f at z_1. We have

$$f(z) = \frac{z}{i\frac{b^2}{4}(z - z_1)^2(z - z_2)^2},$$

so the residue we are looking for is equal to the coefficient of the term $z - z_1$ in the power series expansion of

$$h(z) = \frac{4z}{ib^2(z - z_2)^2}.$$

Differentiating h once we find

$$h'(z) = \frac{4}{ib^2}\left[\frac{-z - z_2}{(z - z_2)^3}\right]$$

which evaluated at z_1 gives

$$\frac{4}{ib^2}\left[\frac{2a/b}{8(\sqrt{a^2 - b^2})^3/b^3}\right] = \frac{a}{i(a^2 - b^2)^{3/2}},$$

which is the residue of f at z_1. Thus

$$\int_C f(z)dz = 2\pi i \frac{a}{i(a^2 - b^2)^{3/2}} = \frac{2\pi a}{(a^2 - b^2)^{3/2}}.$$

Exercise VI.2.24. *Let n be an even integer. Find*

$$\int_0^{2\pi} (\cos\theta)^n d\theta$$

by the method of residues.

Solution. We apply Theorem 2.3 with

$$f(z) = \frac{1}{2^n i z} \left(z + \frac{1}{z} \right)^n.$$

The only pole of f is at the origin. To find the residue of f at 0, we must find the constant term of $\left(z + \frac{1}{z} \right)^n$. Since n is even, the constant term is given by the binomial coefficient

$$\binom{n}{n/2} = \frac{n!}{(n/2)!(n - n/2)!} = \frac{n!}{(n/2)!^2},$$

and therefore, the residue of f at 0 is

$$\frac{n!}{2^n i (n/2)!^2}.$$

Hence

$$\int_0^{2\pi} (\cos \theta)^n d\theta = 2\pi i \frac{n!}{2^n i (n/2)!^2} = \frac{2\pi n!}{2^n (n/2)!^2}.$$

VII
Conformal Mappings

VII.2 Analytic Automorphisms of the Disc

Exercise VII.2.1. *Let f be analytic on the unit disc D, and assume that $|f(z)| < 1$ on the disc. Prove that if there exist two distinct points a, b in the disc which are fixed points, that is, $f(a) = a$ and $f(b) = b$, then $f(z) = z$.*

Solution. Since $|f(z)| < 1$ we have $f(D) \subset D$. Consider the analytic automorphism of the unit disc defined by

$$g(z) = \frac{a - z}{1 - \overline{a}z},$$

and define an analytic function h on the unit disc by $h = g \circ f \circ g$. Then $h(0) = 0$. There exists $w \in D$ such that $g(w) = b$, and since g is its own inverse we find $h(w) = w$. Since a and b are distinct $w \neq 0$. So $|h(w)| = |w|$ with $w \neq 0$. By the Schwarz lemma, there exists a complex number α of absolute value 1 such that $h(z) = \alpha z$. Since $h(w) = w$ we have $\alpha = 1$ and therefore

$$h(z) = \frac{a - f(g(z))}{1 - \overline{a}f(g(z))} = z,$$

and hence $f(g(z)) = g(z)$. Since g is an automorphism of the unit disc we conclude that $f(z) = z$.

Exercise VII.2.2 (Schwarz–Pick Lemma). *Let $f : D \to D$ be a holomorphic map of the disc into itself. Prove that for all $a \in D$ we have*

$$\frac{|f'(a)|}{1 - |f(a)|^2} \leq \frac{1}{1 - a^2}.$$

[Hint: Let g be an automorphism of D such that g(0) = a, and let h be an automorphism which maps f(a) on 0. Let F = h ∘ f ∘ g. Compute F'(0) and apply the Schwarz lemma.]

Solution. Fix $a \in D$, and consider g, h the automorphisms of the disc defined by

$$g(z) = \frac{a - z}{1 - \bar{a}z} \text{ and } h(z) = \frac{f(a) - z}{1 - \overline{f(a)}z}.$$

Define $F : D \to D$ by $F(z) = h \circ f \circ g(z)$. By the chain rule we get

$$F'(0) = h'(f(a))f'(a)g'(0).$$

Direct computations show that $g'(0) = -1 + |a|^2$ and

$$h'(f(a)) = \frac{-1}{1 - |f(a)|^2}.$$

Therefore

$$F'(0) = \frac{f'(a)\left(|a|^2 - 1\right)}{1 - |f(a)|^2}.$$

Since $F(0) = 0$ we can apply the Schwarz lemma which states that $|F'(0)| \leq 1$. The desired inequality drops out.

Exercise VII.2.3. *Let α be a complex number, and let h be an isomorphism of the disc $D(\alpha, R)$ with the unit disc such that $h(z_0) = 0$. Show that*

$$h(z) = \frac{R(z - z_0)}{R^2 - (z - \alpha)(\bar{z}_0 - \bar{\alpha})}e^{i\theta}$$

for some real number θ.

Solution. Consider the isomorphism $g : D(0, 1) \to D(\alpha, R)$ defined by $g(w) = Rw + \alpha$. Let $f = h \circ g$. Then f is an automorphism of the unit disc so by Theorem 2.1 there exists a real number θ such that

$$f(z) = \frac{z_1 - z}{1 - \bar{z}_1 z}e^{i\theta},$$

where $f(z_1) = 0$. We see that $g^{-1}(w) = (w - \alpha)/R$ and $z_1 = g^{-1}(z_0)$. Making the necessary substitutions we get

$$h(z) = f(g^{-1}(z)) = \frac{\frac{z_0 - \alpha}{R} - \frac{z - \alpha}{R}}{1 - \frac{\bar{z}_0 - \bar{\alpha}}{R}\frac{z - \alpha}{R}}e^{i\theta} = \frac{R(z - z_0)}{R^2 - (z - \alpha)(\bar{z}_0 - \bar{\alpha})}e^{i\theta}.$$

Conclude.

Exercise VII.2.4. *What is the image of the half strips as shown on the figure, under the mapping $z \mapsto iz$? Under the mapping $z \mapsto -iz$?*

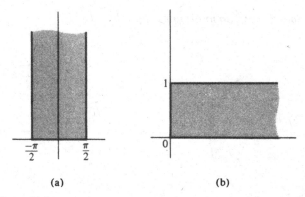

(a)   (b)

Solution. The strip in (a) can be described by $z = x + iy$ with $-\pi/2 \le x \le \pi/2$ and $y \ge 0$, and the strip in (b) is described by $z = x + iy$ with $x \ge 0$ and $0 \le y \le 1$. Since $i(x + iy) = ix - y$ and $-i(x + iy) = -ix + y$ we find that the image of the strip in (a) under the first map (which is a rotation of $\pi/2$) is the strip

$$\{ix - y : -\pi/2 \le x \le \pi/2, \ y \ge 0\},$$

and the image under the second map (which is a rotation of $-\pi/2$) is

$$\{-ix + y : -\pi/2 \le x \le \pi/2, \ y \ge 0\}.$$

The image of the strip in (b) under the first map is the strip

$$\{ix - y : x \ge 0, \ 0 \ge y \ge 1\},$$

and the image of the strip under the second map is

$$\{-ix + y : x \ge 0, \ 0 \ge y \ge 1\}.$$

Exercise VII.2.5. *Let α be real, $0 \le \alpha < 1$. Let U_α be the open set obtained from the unit disc by deleting the segment $[\alpha, 1]$, as shown on the figure.*
(a) Find an isomorphism of U_α with the unit disc from which the segment $[0, 1]$ has been deleted.

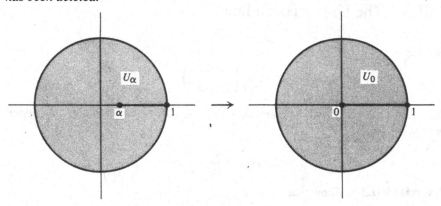

(b) Find an isomorphism of U_0 with the upper half plane of the disc. Also find an isomorphism of U_α with this upper half disc.

[Hint: What does $z \mapsto z^2$ do to the upper half disc?]

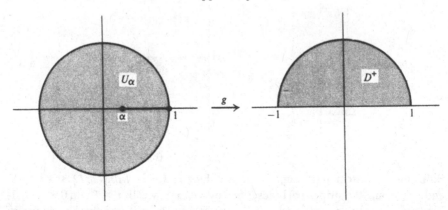

Solution. (a) Consider the function f on the unit disc defined by $f(z) = (z - \alpha)/(1 - \bar{\alpha}z)$. We know that f is a automorphism of the unit disc and since α is real we have $f(z) = (z - \alpha)/(1 - \alpha z)$. We claim that the segment $[\alpha, 1]$ is mapped onto $[0, 1]$. Differentiating $f(z) = (x - \alpha)/(1 - \alpha x)$ we find

$$f'(x) = \frac{1 - \alpha^2}{(1 - \alpha x)^2} > 0$$

and $f(\alpha) = 0$ and $f(1) = 1$, so this proves our claim. Hence f maps U_α onto U_0 isomorphically.

(b) Since U_0 is open and simply connected we can define $g(z) = \sqrt{z} = e^{\frac{1}{2}\log z}$. If $z = re^{i\theta}$, then $g(re^{i\theta}) = e^{\frac{1}{2}(\log r + i\theta)} = \sqrt{r}e^{i\theta/2}$ and it is clear from this that g solves our problem. Composing f and g we get an isomorphism of U_α with the upper half disc.

VII.3   The Upper Half Plane

Let

$$M = \begin{pmatrix} a & b \\ c & d \end{pmatrix}$$

be a 2×2 matrix of real numbers, such that $ad - bc > 0$. For $z \in H$, the upper half plane, define

$$f_M(z) = \frac{az + b}{cz + d}.$$

Exercise VII.3.1. *Show that*

$$\text{Im } f_M(z) = \frac{(ad - bc)y}{|cz + d|^2}.$$

Solution. Writing $z = x + iy$, we see that

$$f_M(z) = \frac{(az + b)(c\bar{z} + d)}{|cz + d|^2} = \frac{(ax + b + iya)(cx + d - iyc)}{|cz + d|^2}.$$

Thus

$$|cz + d|^2 \operatorname{Im} f_M(z) = ya(cx + d) - yc(ax + b) = y(ad - bc).$$

Exercise VII.3.2. *Show that f_M gives a map of H into H.*

Solution. If $z \in H$ and $z = x + iy$, then $y > 0$ and since $ad - bc > 0$ it follows from the previous exercise that $\operatorname{Im} f_M(z) > 0$, whence $f_M(z) \in H$.

Exercise VII.3.3. *Let $\mathrm{GL}_2^+(\mathbf{R})$ denote the set of all 2×2 matrices with positive determinant. Then $\mathrm{GL}_2^+(\mathbf{R})$ is closed under multiplication and taking multiplicative inverses, so $\mathrm{GL}_2^+(\mathbf{R})$ is called a group. Show that if $M, M' \in \mathrm{GL}_2^+(\mathbf{R})$, then*

$$f_{MM'} = f_M \circ f_{M'}.$$

This is verified by brute force. Then verify that if I is the unit matrix,

$$f_I = \mathrm{id} \quad and \quad f_{M^{-1}} = (f_M)^{-1}.$$

Thus every analytic map f_M of H has an analytic inverse, actually in $\mathrm{GL}_2^+(\mathbf{R})$, and in particular f_M is an automorphism of H.

Solution. A straight forward brute force calculation shows that

$$f_{MM'} = f_M \circ f_{M'}.$$

We omit this calculation which follows directly from the definitions. See Exercise VII.5.2. Therefore $\mathrm{id} = f_I = f_{MM^{-1}} = f_M f_{M^{-1}}$ and $f_{M^{-1}} = (f_M)^{-1}$.

Exercise VII.3.4. *(a) If $c \in \mathbf{R}$ and cM is the usual scalar multiplication of a matrix by a number, show that $f_{cM} = f_M$. In particular, let $\mathrm{SL}_2(\mathbf{R})$ denote the subset of $\mathrm{GL}_2^+(\mathbf{R})$ consisting of the matrices with determinant 1. Then given $M \in \mathrm{GL}_2^+(\mathbf{R})$, one can find $c > 0$ such that $cM \in \mathrm{SL}_2(\mathbf{R})$. Hence as far as studying analytic automorphisms of H are concerned, we may concern ourselves only with $\mathrm{SL}_2(\mathbf{R})$. (b) Conversely, show that if $f_M = f_{M'}$ for $M, M' \in \mathrm{SL}_2(\mathbf{R})$, then*

$$M' = \pm M.$$

Solution. (a) It is clear that the c's in the numerator cancel with those in the denominator, so $f_{cM} = f_M$. Also, $\det(cM) = c^2 \det(M)$, so it is clear that given any $M \in \mathrm{GL}_2^+(\mathbf{R})$ there exists $c > 0$ such that $cM \in \mathrm{SL}_2(\mathbf{R})$.
(b) Let $N = M'M^{-1}$. By the above results, we see that if $f_M = f_{M'}$, then $f_N = \mathrm{id}$. If

$$N = \begin{pmatrix} a & b \\ c & d \end{pmatrix}$$

then the above equation gives us

$$\frac{az + b}{cz + d} = z.$$

Clearly, $c = 0$, and evaluating at $z = -b/a$ we find that $b = 0$. Therefore $a = d$ and we find that N is a multiple of the identity. Since $N \in SL_2(\mathbf{R})$ we conclude that $N = \pm I$ and we are done.

Exercise VII.3.5. *(a) Given an element $z = x + iy \in H$, show that there exists an element $M \in SL_2(\mathbf{R})$ such that $f_M(i) = z$.*
*(b) Given $z_1, z_2 \in H$, show that there exists $M \in SL_2(\mathbf{R})$ such that $f_M(z_1) = z_2$. In light of (b), one then says that $SL_2(\mathbf{R})$ acts **transitively** on H.*

Solution. (a) By what was shown above, it suffices to show that there exists an element M in $GL_2^+(\mathbf{R})$ such that $f_M(z) = i$. This is done in two steps. First, a matrix of the form

$$\begin{pmatrix} 1 & b \\ 0 & 1 \end{pmatrix}$$

corresponds to a translation by b. So we may translate z to a point on the imaginary axis, say ri with $r > 0$. Then the matrix

$$\begin{pmatrix} 1 & 0 \\ 0 & r \end{pmatrix}$$

maps ri to i and we are done.
(b) Map z_1 to i and then i to z_2.

Exercise VII.3.6. *Let K denote the subset of elements $M \in SL_2(\mathbf{R})$ such that $f_M(i) = i$. Show that if $M \in K$, then there exists a real θ such that*

$$M = \begin{pmatrix} \cos\theta & -\sin\theta \\ \sin\theta & \cos\theta \end{pmatrix}.$$

Solution. Suppose that

$$\frac{ai + b}{ci + d} = i$$

and $ad - bc = 1$. Then we have

$$ai + b = i(ci + d)$$

so $a = d$ and $b = -c$. Therefore $a^2 + b^2 = 1$, so (a, b) lies on the unit circle, and we conclude that there exists θ such that $a = \cos\theta$ and $c = \sin\theta$.

All Automorphisms of the Upper Half Plane

Do the following exercises after you have read the beginning of §5. In particular, note that Exercise 3 generalizes to fractional linear maps. Indeed, if M, M' denote any complex nonsingular 2×2 matrix, and F_M, $F_{M'}$ are the corresponding fractional linear maps, then

$$F_{MM'} = F_M \circ F_{M'}.$$

Hence if I is the unit 2 × 2 matrix, then

$$F_I = \text{id} \quad \text{and} \quad F_{M^{-1}} = F_M^{-1}.$$

Exercise VII.3.7. *Let* $f : H \to D$ *be the isomorphism of the text, that is*

$$f(z) = \frac{z - i}{z + i}.$$

Note that f *is represented as a fractional linear map,* $f = F_M$ *where* M *is the matrix*

$$M = \begin{pmatrix} 1 & -i \\ 1 & i \end{pmatrix}.$$

Of course, this matrix does not have determinant 1.

Let K *be the set of Exercise 6. Let* $\text{Rot}(D)$ *denote the set of rotations of the unit disc, i.e.,* $\text{Rot}(D)$ *consists of all automorphisms*

$$R_\theta : w \mapsto e^{i\theta} w \quad \text{for } w \in D.$$

Show that $f K f^{-1} = \text{Rot}(D)$, *meaning that* $\text{Rot}(D)$ *consists of all elements* $f \circ f_M \circ f^{-1}$ *with* $M \in K$.

Solution. We compute

$$
\begin{aligned}
f \circ f_M(z) &= \frac{\frac{(\cos\theta)z - \sin\theta}{(\sin\theta)z + \cos\theta} - i}{\frac{(\cos\theta)z - \sin\theta}{(\sin\theta)z + \cos\theta} + i} \\
&= \frac{(\cos\theta - i\sin\theta)z - i(\cos\theta - i\sin\theta)}{(\cos\theta + i\sin\theta)z + i(\cos\theta + i\sin\theta)} \\
&= e^{-2i\theta} f(z).
\end{aligned}
$$

Conclude.

Exercise VII.3.8. *Finally, prove the theorem:*

Theorem. *Every automorphism of* H *is of the form* f_M *for some* $M \in SL_2(\mathbf{R})$.

[Hint: Proceed as follows. Let $g \in \text{Aut}(H)$. *Then there exists* $M \in SL_2(\mathbf{R})$ *such that*

$$f_M(g(i)) = i.$$

By Exercise 6, we have $f_M \circ g \in K$, *say* $f_M \circ g = h \in K$, *and therefore*

$$g = f_M^{-1} \circ h \in SL_2(\mathbf{R}),$$

thus concluding the proof.]

Solution. The existence of f_M follows from Exercise 5(b). Then $f \circ h \circ f^{-1}$ is an automorphism of the disc which fixes 0, i.e., a rotation. So

$$f \circ h \circ f^{-1} \in \text{Rot}(D).$$

Hence $h \in K$ by Exercise 7.

From the Upper Half Plane to the Punctured Disc

Exercise VII.3.9. *Let $f(z) = e^{2\pi i z}$. Show that f maps the upper half plane on the inside of a disc from which the center has been deleted. Given $B > 0$, let $H(B)$ be that part of the upper half plane consisting of those complex number $x + iy$ with $y \geq B$. What is the image of $H(B)$ under f? Is f an isomorphism? Why? How would you restrict the domain of definition of f to make it an isomorphism?*

Solution. If $z = x + iy$, then $e^{2\pi i z} = e^{-2\pi y} e^{2\pi i x}$ and if $y > 0$ we get $0 < |f(z)| = e^{-2\pi y} < 1$. Hence f maps the upper half plane in the interior of the unit disc from which the origin has been deleted. If $f \in H(B)$, then $0 < |f(z)| = e^{-2\pi y} < e^{-2\pi B}$, so the image of $H(B)$ under f is the closed disc centered at the origin of radius $e^{-2\pi B}$ and whose center has been deleted. The function f is not an isomorphism when defined on the upper half plane (or $H(B)$) because z and $z + 1$ have the same image. If we let $S = \{x + iy : 0 \leq x < 1\}$ then the restriction of f to $S \cap H$ or $S \cap H(B)$ is an isomorphism with its image.

VII.4   Other Examples

Exercise VII.4.1. *(a) In each one of the examples, prove that the stated mapping is an isomorphism on the figures as shown. Also determine what the mapping does to the boundary lines. Thick lines should correspond to each other.*
(b) In Example 10, give the explicit formula giving an isomorphism of the strip containing a vertical obstacle with the right half plane, and also with the upper half plane. Note that the counterclockwise rotation by $\pi/2$ is given by multiplication with i.

Solution. (a) **Example 1.** The first quadrant is described by those complex numbers $z = re^{i\theta}$ where $r > 0$ and $0 < \theta < \pi/2$. Then

$$z^2 = r^2 e^{i2\theta}$$

and we see that the image of the first quadrant under the map $f : z \mapsto z^2$ is the upper half plane. The boundary of the first quadrant is given by $\{r = 0\} \cup \{\theta = 0\} \cup \{\theta = \pi/2\}$, and it is clear that its image under f corresponds to the boundary of the upper half plane, namely the real line.

Example 2. The quarter disc corresponds to those complex numbers $z = re^{i\theta}$ with $0 < r < 1$ and $0 < \theta < \pi/2$. It is clear that its image under $f : z \mapsto z^2$ is the half disc. The boundary of the quarter disc is

$$\{r = 0\} \cup \{\theta = 0, \ 0 < r < 1\} \cup \{\theta = \pi/2, 0 < r < 1\} \cup \{|r| = 1\}$$

and it is clear that this set maps to the boundary of the half disc under f.

Example 3. One argues like for the isomorphism between the disc and the upper half plane. If $f(z) = (1+z)/(1-z)$ and $z = x + iy$, then

$$f(z) = \frac{1-(x^2+y^2)}{(1-x)^2+y^2} + i\frac{2y}{(1-x)^2+y^2}$$

so it is clear that f maps the half disc into the first quadrant. The map, defined by $g(w) = \frac{w-1}{w+1}$ on the first quadrant, maps into the half disc, and one easily checks that f and g are inverses of each other. This proves that f is an isomorphism of the half disc with the first quadrant.

If $z = e^{i\theta}$, then

$$f(z) = \frac{1+e^{i\theta}}{1-e^{i\theta}} = \frac{e^{-i\theta/2}+e^{i\theta/2}}{e^{-i\theta/2}-e^{i\theta/2}} = \frac{i}{2\tan(\theta/2)}$$

and we see that the image of the upper half circle is mapped onto the positive imaginary axis. If $z = x$ is real, then

$$f(z) = \frac{1+x}{1-x}$$

and real variable techniques show that f maps the segment $(-1, 1)$ onto the positive real axis.

Example 4. Since the map of this example is obtained from composing maps from previous examples, there is nothing left to prove.

Example 5. If $z = re^{i\theta}$ with $0 < \theta < \pi$ and $0 < r < 1$, then

$$\log z = \log r + i\theta$$

and we see that the real part of $\log z$ is negative and that its imaginary part lies between 0 and π. In fact, $\log : (0, 1) \to (-\infty, 0)$ is a bijection, so we conclude that $\log z$ is an isomorphism of the indicated regions.

To see what happens to the boundary, we first look at $r = 1$. Then $\log z = i\theta$ which shows that $\log$ maps the half circle to the imaginary segment between 0 and $i\pi$. If $\theta = 0$, then $\log z = \log r$ and $\log$ maps the segment $(0, 1]$ bijectively onto the line $(-\infty, 0]$. Finally, if $\theta = \pi$ then $\log z = \log r + i\pi$ and we obtain the last part of the boundary of the strip.

Example 6. This example is very much like the preceding one except that we do not have the restriction $r < 1$. The proof is identical and the details are left to the reader.

Example 7. Again, this example is almost the same as Example 5 except that we do not have the restriction $r < 1$, and that we take $0 < \theta < 2\pi$. The details are left to the reader.

Example 8. If $z = x + iy$, then $e^z = e^x e^{iy}$ so the absolute value of e^z is e^x and its argument is y. It is now clear that $f(z) = e^z$ maps bijectively from the first region onto the second. If $y = 0$, then f maps the real line onto the positive real axis because $x \to e^x$ does so, and if $y = \pi$, then $e^{iy} = -1$ and f maps the line

$x + i\pi$ onto the negative real axis. Finally, if $z = iy$ with $0 \le y \le a$, then $e^z = e^{iy}$ describes the arc contained in the half unit circle from 0 to e^{ia}.

Example 9. From Exercise 3 of the previous section, we see that f is an automorphism of the upper half plane, so the point of interest is to see what happens to the arc $z = e^{i\theta}$ with $0 \le \theta \le a$. In this case, we have

$$
\begin{aligned}
f(e^{i\theta}) &= \frac{e^{i\theta} - 1}{e^{i\theta} + 1} \\
&= \frac{e^{i\theta/2} - e^{-i\theta/2}}{e^{i\theta/2} + e^{-i\theta/2}} \\
&= \frac{2i \sin(\theta/2)}{2 \cos(\theta/2)} \\
&= i \tan(\theta/2).
\end{aligned}
$$

From this equation, the behavior of the isomorphism at the boundary is clear.

Example 10. This is a composite of isomorphisms which are familiar. The square function maps the domain to the plane minus the real number ≤ 1. The translation moves this domain to the plane minus the reals ≤ 0, and the square root function transforms this last domain in the right half plane.

The explanations for Examples 11 and 12 are given in Lang's book.
(b) An isomorphism of the strip containing a vertical obstacle with the right half plane is given by $z \mapsto \sqrt{(-iz)^2 - 1}$. An isomorphism of the strip containing a vertical obstacle with the upper half plane is given by $z \mapsto i\sqrt{(-iz)^2 - 1}$.

Exercise VII.4.2. (a) Show that the function $z \mapsto z + 1/z$ is an analytic isomorphism of the region outside the unit circle onto the plane from which the segment $[-2, 2]$ has been deleted.
(b) What is the image of the unit circle under this mapping? Use polar coordinates.

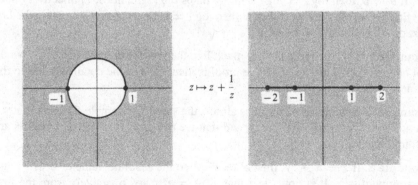

(c) In polar coordinates, if $w = z + 1/z = u + iv$, then

$$u = \left(r + \frac{1}{r}\right)\cos\theta \text{ and } v = \left(r - \frac{1}{r}\right)\sin\theta.$$

Show that the circle $r = c$ with $c > 1$ maps to an ellipse with major axis $c + 1/c$ and minor axis $c - 1/c$. Show that the radial lines $\theta = c$ map onto quarters of hyperbolas.

Solution. (a) Let $z = x + iy$. Then

$$f(z) = x\left(1 + \frac{1}{x^2 + y^2}\right) + iy\left(1 - \frac{1}{x^2 + y^2}\right).$$

Since $x^2 + y^2 > 1$, $\operatorname{Im} f(z) = 0$ if and only if $y = 0$. We now investigate the function of a real variable $f(x)$. Since $f'(x) = 1 - 1/x^2$ the graph of f is

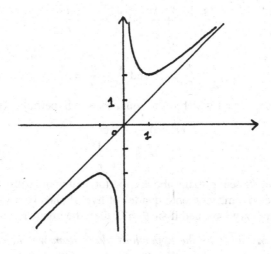

hence

$$f(\mathbf{C} - D) \subset \mathbf{C} - [-2, 2].$$

Suppose $w \in \mathbf{C} - [-2, 2]$. Then $f(z) = w$ is equivalent to

$$z^2 - wz + 1 = 0.$$

Let z_1 and z_2 be the two roots of this equation. We have $z_1 z_2 = 1$ and therefore $|z_1| |z_2| = 1$, so either both roots are on the unit circle or one of them is inside the unit disc and the other is outside. Suppose we are in the first case. Then $z_1 = e^{i\theta}$ and hence $f(z_1) = 2\cos\theta = w$ which is impossible because $w \in \mathbf{C} - [-2, 2]$. So one and only one of the roots is outside the unit disc, which proves that f is injective and surjective.

(b) We use the expression $z = e^{i\theta}$ when z is on the unit circle. Then $f(z) = 2 \cos \theta$ and it is therefore clear that the image of the unit circle is the segment $[-2, 2]$.
(c) If z belongs to the circle $r = c$, then the real and imaginary part of $f(z)$ are

$$x(z) = \left(c + \frac{1}{c}\right) \cos \theta \text{ and } y(z) = \left(c - \frac{1}{c}\right) \sin \theta$$

respectively. When θ ranges over $[0, 2\pi)$ the above is a parametrization of the ellipse centered at the origin with major axis $c + 1/c$ and minor axis $c - 1/c$. Clearly we have

$$\frac{x(z)^2}{\left(c + \frac{1}{c}\right)^2} + \frac{y(z)^2}{\left(c - \frac{1}{c}\right)^2} = 1.$$

Now we show that the radial lines are mapped onto quarter of hyperbolas. Suppose $\theta = c$, $\cos \theta \neq 0$ and $\sin \theta \neq 0$. Let $a = \cos \theta$ and $b = \sin \theta$. Then

$$\frac{u}{a} = r + \frac{1}{r} \text{ and } \frac{v}{b} = r - \frac{1}{r}$$

so that

$$\frac{u}{a} + \frac{v}{b} = 2r \text{ and } \frac{u}{a} - \frac{v}{b} = \frac{1}{2r}$$

hence $u^2/a^2 - v^2/b^2 = 1$ which is the equation of a hyperbola. The map

$$(1, \infty) \to (0, \infty)$$
$$r \mapsto r - \frac{1}{r}$$

is a bijection, and we see from the above equation that u and v are of constant sign so the radial lines are mapped onto quarter of hyperbolas. If $\cos \theta = 0$, then the image is the imaginary axis, and if $\sin \theta = 0$, then the image is $(2, \infty)$.

Exercise VII.4.3. *Let U be the upper half plane from which the points of the closed unit disc are removed, i.e., U is the set of z such that $\mathrm{Im}(z) > 0$ and $|z| > 1$. Give an explicit isomorphism of U with the upper half disc D^+ (the set of z such that $|z| < 1$ and $\mathrm{Im}(z) > 0$.)*

Solution. We compose the inversion map with a rotation, so we let $f(z) = -1/z$. If $z \in U$, then $|z| > 1$, and therefore $|f(z)| < 1$. To see that the image of f belongs to D^+ we write

$$f(x + iy) = -\frac{x - iy}{x^2 + y^2} + i\frac{y}{x^2 + y^2}$$

and note that $\mathrm{Im}\, f(z) > 0$ if and only if $y > 0$. So $f(U) \subset D^+$. Clearly $f(f(z)) = z$ so we conclude that f achieves the desired goal.

Exercise VII.4.4. *Let a be a real number. Let U be the open set obtained from the complex plane by deleting the infinite segment $[a, \infty[$. Find explicitly an analytic isomorphism of U with the unit disc. Give this isomorphism as a composite of* *simpler ones. [Hint: Try first to see what $\sqrt{z}$ does to the set obtained by deleting $[0, \infty[$ from the plane.]*

Solution. The following sequence of transformations shows how to find an analytic isomorphism of U with the unit disc.

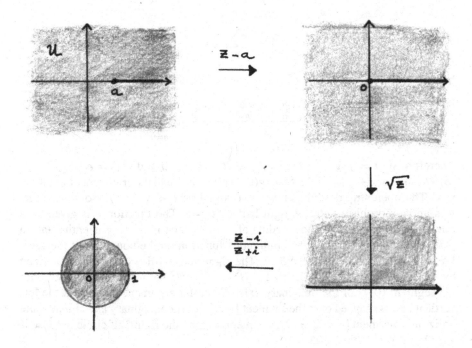

So we can choose

$$f(z) = \frac{\sqrt{z-a}-i}{\sqrt{z-a}+i}.$$

Exercise VII.4.5. *(a) Show that the function $w = \sin z$ can be decomposed as the composite of two functions:*

$$w = \frac{\zeta - \zeta^{-1}}{2i} \text{ and } \zeta = e^{iz} = g(z).$$

(b) Let U be the open half strip in Example 12. Let $g(U) = V$. Describe V explicitly and show that $g : U \to V$ is an isomorphism. Show that g extends to a continuous function on the boundary of U and describe explicitly the image of this boundary under g.

(c) Let $W = f(V)$. Describe W explicitly and show that $f : V \to W$ is an isomorphism. Again describe how f extends continuously to the boundary of V and describe explicitly the image of this boundary under f.

Solution. (a) By definition

$$\sin z = \sum_{n=0}^{\infty} (-1)^n \frac{z^{2n+1}}{(2n+1)!}$$

and

$$e^z = \sum_{n=0}^{\infty} \frac{z^n}{n!}$$

hence

$$\frac{e^{iz} - e^{-iz}}{2i} = \frac{\sum_{n=0}^{\infty} i^n \frac{z^n}{n!} + \sum_{n=0}^{\infty} (-i)^n \frac{z^n}{n!}}{2i} = \sum_{n=0}^{\infty} (-1)^n \frac{z^{2n+1}}{(2n+1)!}.$$

Therefore $\sin z = f(g(z))$ where $f(\zeta) = (\zeta + \zeta^{-1})/2i$ and $g(z) = e^{iz}$.

(b) We have $g(x+iy) = e^{-y}e^{ix}$, so $|g(x+iy)| = e^{-y}$ and the argument of $g(x+iy)$ is x. The open strip described by $y > 0$ and $-\pi/2 < x < \pi/2$ so we see that g is an isomorphism with the right half unit disc. The function g is given by a power series expansion whose radius of convergence is ∞ so g is entire and in particular it can be extend to a continuous function on the boundary of the strip. We now describe the image $g(\partial U)$. The right vertical line can be parametrized by $\pi/2 + it$ with $t \geq 0$ and $g(\pi/2 + it) = e^{-t}e^{i\pi/2}$ so this line is mapped onto the segment $(0, i]$ of the imaginary axis. A similar argument shows that the left vertical line is mapped onto the segment $[-i, 0)$ of the imaginary axis. Finally, the horizontal segment $[-\pi/2, \pi/2]$ gets mapped onto the right half circle because if $-\pi/2 \leq t \leq \pi/2$ then $g(t) = e^{it}$.

(c) We write

$$f(\zeta) = \frac{\zeta - \zeta^{-1}}{2i} = \frac{-1}{2i} \left[i\zeta + (i\zeta)^{-1} \right].$$

Let $\tilde{f}(\zeta) = \zeta + \zeta^{-1}$ so $f(\zeta) = -\tilde{f}(i\zeta)/2$. The map $\zeta \to i\zeta$ is a rotation of angle $\pi/2$ centered at the origin which maps V isomorphically onto the upper half unit disc D^+. Arguing like in Exercise 2 of this section (choosing the root inside the unit disc) we see that $\tilde{f}$ is an analytic isomorphism for D^+ with the lower half plane. Since $f(\zeta) = -\tilde{f}(i\zeta)/2$, it follows that $f : V \to H^+$ is an analytic isomorphism. Note that f is analytic on the complex plane minus the origin. Use the expressions found in Exercise 2 to see that the right half circle of V is mapped onto the segment $[-2, 2]$, and the set $[-i, 0) \cup (0, i]$ is mapped onto $(-\infty, -2] \cup [2, \infty)$.

Exercise VII.4.6. *In Example 12, show that the vertical imaginary axis above the real line is mapped onto itself by $z \mapsto \sin z$, and that this function gives an isomorphism of the half strip with the first quadrant as shown on the figures.*

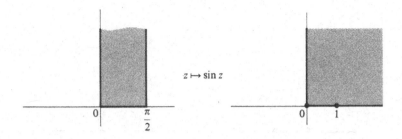

$$z \longmapsto \sin z$$

Solution. We know from Exercise 5 that

$$\sin z = \frac{e^{iz} - e^{-iz}}{2i}$$

so if $z = iy$ with $y > 0$ then

$$\sin iy = \frac{e^{-y} - e^{y}}{2i} = \frac{i}{2} \sinh y,$$

and $\sinh \mathbf{R}_{>0} = \mathbf{R}_{>0}$ so the image of the imaginary axis above the vertical line is mapped into itself by $z \mapsto \sin z$. We use the notation of Exercise 5. From the expression

$$g(x + iy) = e^{-y} e^{ix}$$

we see that g maps the half strip onto the quarter unit disc in the first quadrant. Rotated by $\pi/2$ this quarter disc becomes the unit quarter disc in the second quadrant. The image by $\tilde{f}$ of this quarter unit disc is the third quadrant and this can be seen from the expression of $\tilde{f}$ as a sum of its real and imaginary parts (see Exercise 2 of this section). Since $f(\zeta) = -\tilde{f}(i\zeta)/2$ we get the desired result.

Exercise VII.4.7. *Let $w = u + iv = f(z) = z + \log z$ for z in the upper half plane H. Prove that f gives an isomorphism of H with the open set U obtained from the upper half plane by deleting the infinite half line of numbers*

$$u + i\pi \quad \text{with } u \leq -1.$$

Remark. *The isomorphism f allows us to determine the flow of the lines of a fluid as shown on the figure. These flow lines in the (u, v)-plane correspond to the rays $\theta =$ constant in the (x, y)-plane. In other words, they are the images under f of the rays $\theta =$ constant.*

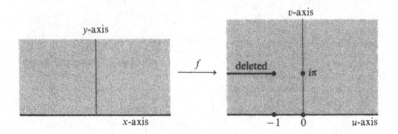

[*Hint: Use Theorem 4.3 applied to the path consisting of the following pieces:*
The segment from R to ϵ (R large, ϵ small > 0).
The small semicircle in the upper half plane, from ϵ to $-\epsilon$.
The segment from $-\epsilon$ to $-R$.
The large semicircle in the upper half plane from $-R$ to R
Note that if we write $z = re^{i\theta}$, then $f(z) = re^{i\theta} + \log r + i\theta.$]

Solution. The path we are considering looks like

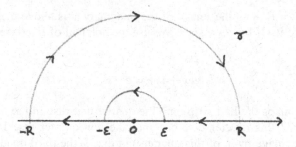

To apply Theorem 4.3 we must investigate the image of the path considered in the hint under f.

Suppose that z is belongs to the segment between ϵ and R. Since $\theta = 0$ we have $f(z) = r + \log r$ which is strictly increasing and gives a bijection of the real line with itself. So the image of the segment is the segment $[\epsilon + \log \epsilon, R + \log R]$ on the real line.

Now suppose that z belongs to the semicircle centered at the origin of radius ϵ contained in the upper half plane. Then we see that $r = \epsilon$ and $0 \leq \theta \leq \pi$ and

$$f(z) = \epsilon \cos \theta + \log \epsilon + i(\theta + \epsilon \sin \theta).$$

Since ϵ is very small we see that the image of this semicircle looks like a vertical segment going from $\epsilon + \log \epsilon$ to $-\epsilon + \log \epsilon + i\pi$.

When z belongs to the real segment $[-R, -\epsilon]$, then $\theta = \pi$ and we get

$$f(z) = -r + \log r + i\pi.$$

Now the graph of the function $g(x) = -r + \log r$ looks like

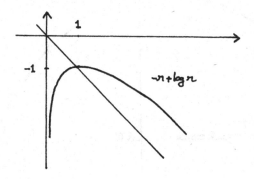

and therefore, we see that the image of the segment $[-R, -\epsilon]$ is the path given by the line segment going from from $-\epsilon + \log \epsilon + i\pi$ to $-1 + i\pi$ and then the line segment going from $-1 + i\pi$ to $-r + \log r + i\pi$.

Finally, on the large semicircle of radius R we have $|z| = R, 0 \le \theta \le \pi$ and

$$f(z) = Re^{i\theta} + \log R + i\theta.$$

Now $Re^{i\theta}$ traces out a semicircle in the upper half plane, which we translate by $\log R$ and add the vertical perturbation given by $i\theta$. We conclude that the image of the path we consider in the upper half plane looks like the figure at the top of the facing page.

If γ denotes the initial path in the upper half plane given by the hint, we see that γ satisfies the hypothesis of Theorem 4.3, both γ and $f \circ \gamma$ have interiors and the interior of $f \circ \gamma$ is connected, so the conclusion of Theorem 4.3 holds. Now let $\epsilon \to 0$ and $R \to \infty$ to conclude the exercise.

Exercise VII.4.8. *Give another proof of Example 11 using Theorem 4.3.*

Solution. Consider the path γ drawn on the figure in the middle of the facing page.

As we did in the previous exercise, we must look at $f \circ \gamma$ where $f(z) = z + 1/z$.

The image of the segment $[1, R]$ is the segment $[2, R + 1/R]$, as one sees after graphing the function $x + 1/x$.

If z belongs to the semicircle of radius 1, then $|z| = 1, 0 \le \theta \le \pi$ and

$$f(z) = e^{i\theta} + e^{-i\theta} = 2\cos\theta.$$

Therefore, the image of this semicircle is the segment $[-2, 2]$.

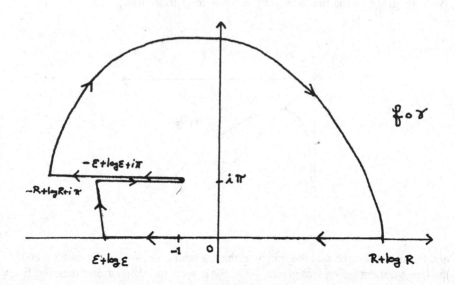

$f \circ \gamma$

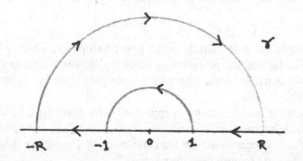

γ

The image of the segment $[-R, 1]$ is the segment $[-R - 1/R, -2]$, as one sees after graphing the function $x + 1/x$.

Finally when z belongs to the large semicircle of radius R we have $|z| = R$, $0 \le \theta \le \pi$ and

$$f(z) = Re^{i\theta} + \frac{1}{Re^{i\theta}}.$$

When R is large, we see from the above expression that the image of the large semicircle is a small perturbation of the semicircle of radius R. We conclude that the image of the path γ looks like

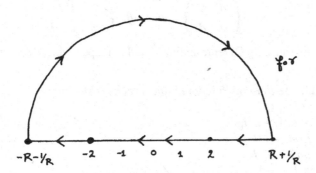

Now we can apply Theorem 4.3 and let $R \to \infty$.

VII.5 Fractional Linear Transformations

Exercise VII.5.1. *Give explicitly a fractional linear map which sends a given complex number z_1 to ∞. What is the simplest such map which sends 0 to ∞?*

Solution. Let

$$\gamma_{z_1} = \begin{pmatrix} 0 & 1 \\ 1 & -z_1 \end{pmatrix} \in GL_2(\mathbf{C}).$$

Then the associated fractional linear map

$$g(z) = \frac{1}{z - z_1}$$

sends z_1 to ∞. The simplest fractional linear map sending 0 to ∞ is the inversion $z \mapsto 1/z$.

Exercise VII.5.2. *Composition of Fractional Linear Maps.* *Show that if F, G are fractional linear maps, then so if $F \circ G$.*

Solution. Let $F(z) = (az + b)/(cz + d)$ and $F(z) = (a'z + b')/(c'z + d')$, then

$$F(G(z)) = \frac{a\left(\frac{a'z+b'}{c'z+d'}\right) + b}{c\left(\frac{a'z+b'}{c'z+d'}\right) + d} = \frac{(aa' + bc')z + (ab' + bd')}{(ca' + dc')z + (b'c + dd')}.$$

Some simple computations show that

$$(aa' + bc')(b'c + dd') - (ca' + dc')(ab' + bd') = (ad - bc)(a'b' - d'c') \neq 0$$

and therefore $F \circ G$ is a fractional linear map. Note that if

$$\alpha = \begin{pmatrix} a & b \\ c & d \end{pmatrix} \quad \text{and} \quad \beta = \begin{pmatrix} a' & b' \\ c' & d' \end{pmatrix}$$

are two matrices in $GL_2(\mathbf{C})$ representing F and G respectively, then $\alpha\beta \in GL_2(\mathbf{C})$ represents $F \circ G$.

Exercise VII.5.3. *Find the fractional linear maps which map:*
(a)$1, i, -1$ on $i, -1, 1$
(b) $i, -1, 1$ on $-1, -i, 1$
(c) $-1, -i, 1$ on $-1, 0, 1$
(d) $-1, 0, 1$ on $-1, i, 1$
(e)$1, -1, i$ on $1, i, -1$

Solution. (a) $F(z) = \frac{(1+2i)z+1}{z+(1-2i)}$

(b)$F(z) = \frac{(2i-1)z+1}{z+(1+2i)}$

(c)$F(z) = \frac{z+i}{iz+1}$

(d)$F(z) = \frac{z+i}{iz+1}$

(e)$F(z) = \frac{(i-1)z+3+i}{(1+3i)z+1-i}$

Exercise VII.5.4. *Find the fractional linear maps which map:*
(a) $0, 1, \infty$ on $1, \infty, 0$
(b) $0, 1, \infty$ on $-1, -1, i$
(c) $0, 1, \infty$ on $-1, 0, 1$
(d) $0, 1, \infty$ on $-1, -i, 1$

Solution. (a) $F(z) = \frac{1}{-z+1}$

(b) $F(z) = \frac{-2iz+i+1}{-2z+i+1}$

(c) $F(z) = \frac{z-1}{z+1}$

(d) $F(z) = \frac{-iz-1}{-iz+1}$

Exercise VII.5.5. *Let F and G be two fractional linear maps, and assume that $F(z) = G(z)$ for all complex numbers z (or even for three distinct complex numbers z). Show that if*

$$F(z) = \frac{az+b}{cz+d} \quad \text{and} \quad G(z) = \frac{a'z+b'}{c'z+d'}$$

then there exists a complex number λ such that

$$a' = \lambda a, \quad b' = \lambda b, \quad c' = \lambda c, \quad d' = \lambda d.$$

Thus the matrices representing F and G differ by a scalar.

Solution. Let g_γ denote the fractional linear map represented by $\gamma \in GL_2(\mathbf{C})$. We want to show that if $g_\gamma = g_{\gamma'}$ for at least three distinct complex numbers then γ and γ' differ by a scalar. By Exercise 2 we know that $g_\gamma^{-1} = g_{\gamma^{-1}}$ and that $g_{\gamma'} g_\gamma^{-1} = g_{\gamma'\gamma^{-1}}$. Hence $g_{\gamma'\gamma^{-1}}$ has at least three fixed points which implies that $g_{\gamma'\gamma^{-1}}$ is the identity. Thus $\gamma'\gamma^{-1} = \lambda I$ for some complex number λ.

Exercise VII.5.6. *Consider the fractional linear map*

$$F(z) = \frac{z - i}{z + i}.$$

What is the image of the real line **R** *under this map? (You have encountered this map as an isomorphism between the upper half plane and the unit disc.)*

Solution. If x is a real number we can express $F(x)$ as the sum of its real and imaginary parts,

$$F(x) = \frac{x - i}{x + i} = \frac{x^2 - 1}{x^2 + 1} + i\frac{-2x}{x^2 + 1}.$$

We parametrize the real line by $x(t) = \tan t$ with $t \in (\pi/2, \pi/2)$. Then the standard trigonometric identities give

$$\begin{cases} \operatorname{Re}(F(x(t))) = -\cos 2t, \\ \operatorname{Im}(F(x(t))) = -\sin 2t. \end{cases}$$

From this system we see at once that the image of the real line under F is the unit circle from which 1 has been deleted, that is

$$F(\mathbf{R}) = C - \{1\}.$$

If we take the real line $\mathbf{R} \cup \{\infty\}$ on the Riemann sphere, then the image is the whole unit circle.

Exercise VII.5.7. *Let F be the fractional linear map $F(z) = (z - 1)/(z + 1)$. What is the image of the real line under this map? (Cf. Example 9 of §4.)*

Solution. If x is real, then

$$F(x) = \frac{x - 1}{x + 1}$$

so $F(x)$ is also real. Differentiating the quotient we get

$$\frac{d}{dx}\left(\frac{x - 1}{x + 1}\right) = \frac{2}{(x + 1)^2} > 0$$

so viewed as a function of a real variable, F is increasing and continuous everywhere on $\mathbf{R} - \{1\}$. Furthermore the limits

$$\lim_{x \to -\infty} F(x) = \lim_{x \to \infty} F(x) = 1, \quad \lim_{x \to -1^+} F(x) = -\infty, \quad \lim_{x \to -1^-} F(x) = \infty$$

show that F is a bijection of $\mathbf{R} - \{-1\}$ with $\mathbf{R} - \{1\}$.

Exercise VII.5.8. *Let $F(z) = z/(z - 1)$ and $G(z) = 1/(1 - z)$. Show that the set of all possible fractional linear maps which can be obtained by composing F and G above repeatedly with each other in all possible orders in fact has six elements, and give a formula for each one of these. [Hint: Compute F^2, F^3, G^2, G^3, $F \circ G$, $G \circ F$, etc.]*

Solution. Let

$$S = \{F(z), G(z), z, 1/F(z), 1/G(z), 1/z\}.$$

A direct computation shows that $F^2(z) = z$, so $F^{2p} = $ id and $F^{2p+1} = F$ therefore F^n belongs to S for all n. Similarly, $G^2 = 1/F$ and $G^3 = $ id hence $G^{2+3p} = 1/F$, $G^{3+3p} = $ id and $G^{4+3p} = G$, which implies that $G^m \in S$ for all m. Some algebra shows that $FH \in S$ for all $H \in S$ and $GH \in S$ for all $H \in S$ so by induction it is clear that $F^m G^n$ and $G^m F^n$ belong to S for all m and n. This implies that the composites of F and G belong to S because given any expression we look at pairs which reduces the problem to elements of the form $F^m G^n$ or $G^m F^n$ which we know belongs to S.

Exercise VII.5.9. *Let $F(z) = (z - i)/(z + i)$. What is the image under F of the following sets of points:*
(a) The upper half line it, with $t \geq 0$.
(b) The circle of center 1 and radius 1.
(c) The horizontal line $i + t$, with $t \in \mathbf{R}$.
(d) The half circle $|z| = 2$ with $\operatorname{Im} z \geq 0$.
(e) The vertical line $\operatorname{Re} z = 1$ and $\operatorname{Im} z \geq 0$.

Solution. (a) We have

$$F(it) = \frac{t - 1}{t + 1}$$

so we find that the image of the upper half line is the half open segment $[-1, 1)$.
(b) Let C be the circle of radius 1 centered at 1. We can write

$$F(z) = 1 + \frac{-2i}{z + i}.$$

The equation of the circle of radius 1 centered at $1 + i$ is

$$(x - 1)^2 + (y - 1)^2 = 1$$

which is equivalent to $2x + 2y - 1 = x^2 + y^2$. So from the proof of Theorem 5.2 we see that the image of this circle under the inversion is the circle whose equation is $-(u^2 + v^2) + 2u - 2v = 1$, namely the circle centered at $1 - i$ of radius 1. So after a multiplication and a translation we see that the image under of C under F us the circle centered at $-2i - 1$ and of radius 2.
(c) Arguing exactly like in (b) we see that the image of the line $i + t$ under the map $1/(z + i)$ is the circle of radius $1/4$ centered at $-i/4$. So the image of the line $i + t$ under the map F is the circle centered at $1 - 2i(-i/4) = 1/2$ of radius $1/2$.
(d) The equation of the circle centered at the origin and of radius 2 is

$$x^2 + y^2 = 4.$$

Let C^+ be the upper half of this circle. Arguing like in (b) we find that the image of this circle under the map $1/(z + i)$ is the circle K centered at $1 + i/3$ of radius $2/3$. The image of C^+ is therefore the lower arc A on K joining $(-2 - i)/5$ to $(2 - i)/5$. So the image of C^+ under F is $1 - 2iA$.

(e) Arguing like in (b) we find that the image of the line $\text{Re}(z) = 1$ under the map $1/(z+i)$ is the circle C centered at $1/2$ and of radius $1/2$. The image Q of the half line $\text{Re}(z) = 1$ and $\text{Im}(z) \geq 0$ under $1/(z+i)$ is the lower left quarter circle of C including $1/2 + i/2$ but excluding 0. Therefore the image of the line $\text{Re}(z) = 1$ and $\text{Im}(z) \geq 0$ under F is $1 - 2i Q$.

Exercise VII.5.10. *Find fractional linear maps which map:*
(a) 0, 1, 2 to 1, 0, ∞
(b) i, -1, 1 to 1, 0, ∞
(c) 0, 1, 2 to i, -1, 1

Solution. (a) $F(z) = \frac{2z-2}{z-2}$

(b) $F(z) = \frac{iz+i}{z-1}$

(c) $F(z) = \frac{(1+3i)z-4i}{(i+3)z-4}$

Exercise VII.5.11. *Let $F(z) = (z + 1)/(z - 1)$. Describe the image of the line $\text{Re}(z) = c$ for a real number c. (Distinguish $c = 1$ and $c \neq 1$. In the second case, the image is a circle. Give its center and radius.)*

Solution. If $z = c + iy$ then

$$F(c + iy) = \frac{c+1+iy}{c-1+iy}.$$

If $c = 1$, then

$$F(1 + iy) = \frac{2+iy}{iy} = 1 - i\frac{2}{y}$$

so the image of the line $\text{Re}(z) = 1$ is the line $\text{Re}(z) = c$ minus the point $(1, 0)$.

If $c \neq 1$ we must show that the image of the line $\text{Re}(z) = c$ is a circle. To do so, write F as the composite of simpler functions, namely write

$$F(z) = 1 + \frac{2}{z+1}.$$

Let $a = c - 1$. The image of the line $\text{Re}(z) = c$ under the map $1/(z - 1)$ is the image of the line $\text{Re}(z) = a$ under the inversion $1/z$. We have

$$\frac{1}{a+iy} = \frac{a}{a^2 + y^2} + i\frac{-y}{a^2 + y^2} = u + iv$$

and from the proof of Theorem 5.3 we find the relation

$$-au^2 - av^2 + u = 0,$$

or equivalently

$$u^2 + v^2 - \frac{1}{a}u = 0.$$

Completing the square we find

$$\left(u - \frac{1}{2a}\right)^2 + v^2 = \frac{1}{4a^2}.$$

Therefore the image of the line $\text{Re}(z) = a$ under the inversion is the circle centered at $1/(2a)$ and of radius $1/|2a|$. Multiplying by 2 and translating by 1 we find that the image of the line $\text{Re}(z) = c$ under F is the circle centered at $1 + 1/(c-1)$ and whose radius is $1/|c-1|$.

Exercise VII.5.12. *Let z_1, z_2, z_3, z_4 be distinct complex numbers. Define their cross ratio to be*

$$[z_1, z_2, z_3, z_4] = \frac{(z_1 - z_3)(z_2 - z_4)}{(z_2 - z_3)(z_1 - z_4)}.$$

(a) Let F be a fractional linear map. Let $z_i' = F(z_i)$ for $i = 1, \ldots, 4$. Show that the cross ratio of z_1', z_2', z_3', z_4' is the same as the cross ratio of z_1, z_2, z_3, z_4. It will be easy if you do it separately for translations, inversions, and multiplications.
(b) Prove that the four numbers lie on the same straight line or on the same circle if and only if their cross ratio is a real number.
(c) Let z_1, z_2, z_3, z_4 be distinct complex numbers. Assume that they lie on the same circle, in that order. Prove that

$$|z_1 - z_3|\,|z_2 - z_4| = |z_1 - z_2|\,|z_3 - z_4| + |z_2 - z_3|\,|z_4 - z_1|.$$

Solution. We can write F as a composite of translations, multiplications and inversion. It is therefore sufficient to show that the cross ratio is invariant under the above three transformations. For the first two, the assertion is obvious so we only have to look at the inversion. Staying cool calm and collected we find

$$\left[\frac{1}{z_1}, \frac{1}{z_2}, \frac{1}{z_3}, \frac{1}{z_4}\right] = \frac{\left(\frac{1}{z_1} - \frac{1}{z_3}\right) - \left(\frac{1}{z_2} - \frac{1}{z_4}\right)}{\left(\frac{1}{z_2} - \frac{1}{z_3}\right) - \left(\frac{1}{z_1} - \frac{1}{z_4}\right)} = \frac{\left(\frac{z_3 - z_1}{z_1}z_3\right)\left(\frac{z_4 - z_2}{z_2}z_4\right)}{\left(\frac{z_3 - z_2}{z_2}z_3\right)\left(\frac{z_4 - z_1}{z_1}z_4\right)}$$

$$= \frac{(z_3 - z_1)(z_4 - z_2)}{(z_3 - z_2)(z_4 - z_1)} = [z_1, z_2, z_3, z_4]$$

as was to be shown.
(b) Suppose that the four numbers z_1, z_2, z_3, z_4 lie on the same straight line or on the same circle and that these numbers are all pairwise distinct, otherwise their cross ratio is zero and there is nothing to prove. Choose three distinct real numbers, say 0, 1 and 2. There exists a fractional linear map which sends z_1, z_2, z_3 to 0, 1 and 2 respectively. Since the image of a circle or a line is a circle or a line, it follows that the image of z_4 also belongs to the real numbers. If follows from (a) that the cross ratio of z_1, z_2, z_3, z_4 is real.

Conversely, suppose that $[z_1, z_2, z_3, z_4]$ is real. Let F be the fractional linear map which sends z_1, z_2, z_3 to 0, 1 and 2 respectively, and let $z_4' = F(z_4)$. Part (a) implies $[0, 1, 2, z_4']$ is real, hence $z_4' \in \mathbf{R}$. But F^{-1} is a fractional linear map and $z_i \in F^{-1}\mathbf{R}$ for $i = 1, \ldots, 4$ so z_1, z_2, z_3, z_4 belong to the same line or circle.
(c) Let F be the fractional linear map which sends z_1, z_2, z_3 to $1, 2$ and 3 respectively. Then either $F(z_4) < 0$ or $F(z_4) > 3$. In all cases we have

$$[1, 2, 3, z_4'] > 0 \quad \text{and} \quad [1, 3, 2, z_4'] < 0$$

hence

$$[z_1, z_2, z_3, z_4] > 0 \quad \text{and} \quad [z_1, z_3, z_2, z_4] < 0.$$

A direct calculation shows that

$$(z_1 - z_3)(z_2 - z_4) = (z_1 - z_2)(z_3 - z_4) - (z_2 - z_3)(z_4 - z_1),$$

hence

$$-[z_1, z_2, z_3, z_4] = [z_1, z_3, z_2, z_4] - 1.$$

From the fact that $[z_1, z_2, z_3, z_4] > 0$ and $[z_1, z_3, z_2, z_4] < 0$ we get

$$|[z_1, z_2, z_3, z_4]| = |[z_1, z_3, z_2, z_4]| + 1 < 0.$$

Conclude.

Fixed Points and Linear Algebra

Exercise VII.5.13. *Find the fixed points of the following functions:*
(a) $f(z) = \frac{z-3}{z+1}$
(b) $f(z) = \frac{z-4}{z+2}$
(c) $f(z) = \frac{z-i}{z+1}$
(d) $f(z) = \frac{2z-3}{z+1}$

Solution. (a) $i\sqrt{3}$ and $-i\sqrt{3}$.
(b) $\frac{-1+i\sqrt{15}}{2}$ and $\frac{-1-i\sqrt{15}}{2}$.
(c) $e^{-i\pi/4}$ and $e^{i3\pi/4}$.
(d) $\frac{1+i\sqrt{11}}{2}$ and $\frac{1-i\sqrt{11}}{2}$.

Exercise VII.5.14. *Let M be a 2×2 complex matrix with nonzero determinant,*

$$M = \begin{pmatrix} a & b \\ c & d \end{pmatrix}, \quad \text{and } ad - bc \neq 0.$$

Define $M(z) = (az+b)/(cz+d)$ as in the text for $z \neq -d/c$ ($c \neq 0$). If $z = -d/c$ ($c \neq 0$) we put $M(z) = \infty$. We define $M(\infty) = a/c$ if $c \neq 0$, and ∞ if $c = 0$.
(a) If L, M are two complex matrices as above, show directly that

$$L(M(z)) = (LM)(z)$$

for $z \in \mathbf{C}$ or $z = \infty$. Here LM is the product of matrices from linear algebra.
(b) Let λ, λ' be the eigenvalues of M viewed as a linear map on $\mathbf{C}^2$. Let

$$W = \begin{pmatrix} w_1 \\ w_2 \end{pmatrix} \quad \text{and } W' = \begin{pmatrix} w_1' \\ w_2' \end{pmatrix}$$

be the corresponding eigenvectors, so

$$MW = \lambda W \text{ and } MW' = \lambda' W'$$

*By a **fixed point** of M on **C** we mean a complex number z such that M(z) = z. Assume that M has two distinct fixed points in **C**. Show that these fixed points are $w = w_1/w_2$ and $w' = w_1'/w_2'$.*

(c) Assume that $|\lambda| < |\lambda'|$. Given $z \neq w$, show that

$$\lim_{k \to \infty} M^k(z) = w'.$$

Note. The iteration of the fractional linear map is sometimes called a dynamical system. Under the assumption in (c), one says that w' is an attracting point for the map and that w is a repelling point.

Solution. See Exercise 2 of this section for the case when all the quantities belong to **C**. We consider the cases which are left. Let

$$M = \begin{pmatrix} a' & b' \\ c' & d' \end{pmatrix} \text{ and } L = \begin{pmatrix} a & b \\ c & d \end{pmatrix}$$

so that

$$LM = \begin{pmatrix} aa' + bc' & ab' + bd' \\ ca' + dc' & cb' + dd' \end{pmatrix}$$

The different cases are

$c' \neq 0, z = -d'/c'$ and $c \neq 0$
$c' \neq 0, z = -d'/c'$ and $c = 0$
$c' \neq 0, z = \infty$ and $c \neq 0$
$c' \neq 0, z = \infty$ and $c = 0$
$c' = 0$, and $c \neq 0$
$c' = 0$, and $c = 0$

In the first case we have $L(M(z)) = a/c$ and

$$LM(z) = \frac{\frac{-aa'd'}{c'} + ab'}{\frac{-ca'd'}{c'} + cb'} = \frac{a}{c}.$$

The same kind of argument settles the remaining cases.

(b) A fixed point for M satisfies

$$\frac{az + b}{cz + d} = z.$$

So z is a fixed point if and only if

$$az + b = cz^2 + dz.$$

Clearly, $c \neq 0$ because there are two fixed points, and the discriminant of this quadratic is equal to the discriminant of the determinant of $M - tI$ so having two distinct fixed points implies that the eigenvalues are distinct and the eigenvectors are therefore linearly independent. This implies that $w_1/w_2 \neq w_1'/w_2'$. Now $MW = \lambda W$ implies

$$aw_1 + bw_2 = \lambda w_2 \quad \text{and} \quad cw_1 + dw_2 = \lambda w_2.$$

Since M is invertible $\lambda \neq 0$ and $\lambda' \neq 0$. Also, $w_2 \neq 0$ for otherwise, we see from above that $c = 0$ and this is a contradiction. It is clear from the above equations that $M(w_1/w_2) = w_1/w_2$. A similar argument for W' concludes the proof.

(c) Let $\alpha = \lambda/\lambda'$. Since the expression w and w' where defined in (b) we assume that the hypothesis of (b) hold, and in particular $c \neq 0$. Since $\lambda \neq \lambda'$ the argument at the beginning of (b) shows that M has two distinct fixed points given by w and w'. We have also seen that $w_2 \neq 0$ and $w'_2 \neq 0$ so we rescale W and W' and assume that $W = (w, 1)$ and $W' = (w', 1)$. Let S be the matrix

$$S = \begin{pmatrix} w & w' \\ 1 & 1 \end{pmatrix}.$$

This matrix corresponds to the change of variable where the basis are the eigenvectors, so

$$S^{-1}MS = \begin{pmatrix} \lambda & 0 \\ 0 & \lambda' \end{pmatrix}.$$

Therefore

$$S^{-1}M^kS = \begin{pmatrix} \lambda^k & 0 \\ 0 & (\lambda')^k \end{pmatrix}$$

so that $S^{-1}M^kS(z) = \alpha^k z$. Since $z \neq 0$, $S^{-1}(z) \in \mathbf{C}$ and therefore

$$M^k(z) = S(\alpha^k S^{-1}(z)).$$

But $\lim_{k\to\infty} \alpha^k S^{-1}(z) = 0$ which implies

$$\lim_{k\to\infty} M^k(z) = S(0) = w'.$$

This argument can be found in the appendix of Lang's book.

VIII
Harmonic Functions

VIII.1   Definition

Exercise VIII.1.1. *(a) Let* $\Delta = \left(\frac{\partial}{\partial x}\right)^2 + \left(\frac{\partial}{\partial y}\right)^2$. *Verify that*

$$\Delta = 4\frac{\partial}{\partial z}\frac{\partial}{\partial \bar{z}}.$$

(b) Let f be a complex function on $\mathbf{C}$ such that both f and f^2 are harmonic. Show that f is holomorphic or $\bar{f}$ is holomorphic.

Solution. (a) We have

$$\frac{\partial f}{\partial \bar{z}} = \frac{1}{2}\left(\frac{\partial f}{\partial x} + i\frac{\partial f}{\partial y}\right),$$

$$\frac{\partial f}{\partial z}\frac{\partial f}{\partial x} = \frac{1}{2}\left(\frac{\partial^2 f}{\partial x^2} - i\frac{\partial^2 f}{\partial y \partial x}\right),$$

$$\frac{\partial f}{\partial z}\frac{\partial f}{\partial y} = \frac{1}{2}\left(\frac{\partial^2 f}{\partial y \partial x} - i\frac{\partial^2 f}{\partial y^2}\right).$$

The third equation is true because the partials commute. Hence

$$4\frac{\partial f}{\partial z}\frac{\partial f}{\partial \bar{z}} = \frac{\partial^2 f}{\partial x^2} - i\frac{\partial^2 f}{\partial y \partial x} + i\frac{\partial^2 f}{\partial y \partial x} + \frac{\partial^2 f}{\partial y^2} = \Delta f.$$

(b) Since f^2 is harmonic, we have

$$\frac{\partial}{\partial z}\frac{\partial}{\partial \bar{z}}f^2 = 0.$$

Now we can use the product rule to obtain

$$\frac{\partial}{\partial z}\frac{\partial}{\partial \bar{z}}f^2 = 2\frac{\partial f}{\partial z}\frac{\partial f}{\partial \bar{z}} + 2f\frac{\partial}{\partial z}\frac{\partial}{\partial \bar{z}}f.$$

But f is harmonic, so the last term vanishes. We conclude that

$$0 = \frac{\partial f}{\partial z}\frac{\partial f}{\partial \bar{z}}.$$

If $S_1 = \{z \in \mathbf{C} : \partial f/\partial \bar{z} = 0\}$ and $S_2 = \{z \in \mathbf{C} : \partial f/\partial z = 0\}$ then $S_1 \cup S_2 = \mathbf{C}$. If $S_1 = \mathbf{C}$, then f is holomorphic. If $S_1 \neq \mathbf{C}$, then S_2 is nonempty and contains an open ball because S_1 is closed. So the function $u = \partial f/\partial z$ vanishes on an open set. We claim that u is harmonic. Indeed, by (a) we have

$$\Delta u = 4\frac{\partial}{\partial z}\left(\frac{\partial}{\partial \bar{z}}\frac{\partial}{\partial z}f\right),$$

and arguing like in (a) it is easy to show that

$$\Delta = 4\frac{\partial}{\partial \bar{z}}\frac{\partial}{\partial z}.$$

Since f is harmonic, our claim is proved. Finally, let S be the set of points in $\mathbf{C}$ where u vanishes, and let U be its interior, which we have shown to be nonempty. If we can show that U is also closed, then, since $\mathbf{C}$ is connected, we will conclude that $u = 0$ everywhere and therefore, $\bar{f}$ is holomorphic. To prove that U is closed, let $\{w_k\} \subset U$ be a sequence of points converging to w. We want to show that $w \in U$. Since harmonic functions are locally the real parts of analytic functions, choose a neighborhood V of w and an analytic function g defined on V such that $\mathrm{Re}(g) = u$. For some large k, $w_k \in V$ so that u vanishes on a small open set in V. On this open set, g is purely imaginary so by the open mapping theorem, we conclude that g is equal to an imaginary constant on V. Hence u is identically zero on V and $w \in U$. This concludes the proof.

Exercise VIII.1.2. *Let f be analytic, and $\bar{f} = u - iv$ the complex conjugate function. Verify that $\partial \bar{f}/\partial z = 0$.*

Solution. We see that

$$\begin{aligned}
\frac{\partial \bar{f}}{\partial z} &= \frac{1}{2}\left(\frac{\partial \bar{f}}{\partial x} - i\frac{\partial \bar{f}}{\partial y}\right) \\
&= \frac{1}{2}\left(\frac{\partial u}{\partial x} - i\frac{\partial v}{\partial x} - i\frac{\partial u}{\partial y} - \frac{\partial v}{\partial y}\right),
\end{aligned}$$

and therefore

$$\overline{\frac{\partial \bar{f}}{\partial z}} = \frac{1}{2}\left(\frac{\partial u}{\partial x} + i\frac{\partial v}{\partial x} + i\frac{\partial u}{\partial y} - \frac{\partial v}{\partial y}\right) = \frac{\partial f}{\partial \bar{z}}.$$

So we have the general formula

$$\frac{\overline{\partial \overline{f}}}{\partial z} = \frac{\partial f}{\partial \overline{z}}.$$

Since f is analytic if and only if $\partial f/\partial \overline{z} = 0$, we conclude that $\partial \overline{f}/\partial z = 0$ if and only if f is analytic.

Exercise VIII.1.3. *Let $f : U \to V$ be an analytic isomorphism, and let φ be a harmonic function on V, which is the real part of an analytic function. Prove that the composite $\varphi \circ f$ is harmonic.*

Solution. Let g be an analytic function such that

$$g(x + iy) = \varphi(x, y) + i\psi(x, y)$$

where $\varphi, \psi : \mathbf{R}^2 \to \mathbf{R}$ are harmonic. If $f = u + iv$, then

$$g \circ f = \varphi(u(x, y), v(x, y)) + i\psi(u(x, y), v(x, y))$$

and $g \circ f$ is analytic so $\varphi(u(x, y), v(x, y))$ is harmonic as was to be shown.

Exercise VIII.1.4. *Prove that the imaginary part of an analytic function is harmonic.*

Solution. Let f be an analytic function, and let $g = -if$. Then g is analytic and the real part of g is the imaginary part of f. Conclude.

Exercise VIII.1.5. *Prove the uniqueness statement in the following context. Let U be an open set contained in a strip $a \le x \le b$, where a, b are fixed numbers, and as usual $z = x + iy$. Let u be a continuous function on $\overline{U}$, harmonic on U. Assume that u is 0 on the boundary of U, and*

$$\lim u(x, y) = 0$$

as $y \to \infty$ or $y \to -\infty$, uniformly in x. In other words, given ϵ there exists $C > 0$ such that if $y > C$ or $y < -C$ and $(x, y) \in U$ then $|u(x, y)| < \epsilon$. Then $u = 0$ on U.

Solution. If U is bounded, the proof is given in the text, so we assume that U is unbounded. Suppose that u is not identically zero. Then since $u(x, y) \to 0$ as $|y| \to \infty$ uniformly in x, we conclude that u attains a maximum in U, say at (x_0, y_0) which is obviously an interior point of U. Working with $-u$ if necessary we may assume that $u(x_0, y_0) > 0$. Given $\epsilon > 0$ let φ_ϵ be defined by

$$\varphi_\epsilon(x, y) = u(x, y) + \epsilon x^2.$$

Since $|x|$ is bounded, the function φ_ϵ attains a maximum on U. If (x, y) is on the boundary of U, then

$$\varphi_\epsilon(x, y) = \epsilon x^2,$$

and since $|x|$ is bounded, we can select ϵ so small that $\varphi_\epsilon(x, y) < u(x_0, y_0)$ on the boundary of U. Having chosen such an ϵ we see that φ_ϵ does not attain its maximum on a boundary point of U because if $(x, y) \in \partial U$, then

$$\varphi_\epsilon(x, y) = \epsilon x^2 < u(x_0, y_0) + \epsilon x_0^2 = \varphi_\epsilon(x_0, y_0).$$

So φ_ϵ attains its maximum at an interior point of U, say at (x_1, y_1). It follows that

$$D_1^2 \varphi_\epsilon(x_1, y_1) \le 0 \quad \text{and} \quad D_2^2 \varphi_\epsilon(x_1, y_1) \le 0.$$

But since u is harmonic we have $(D_1^2 + D_2^2)u = 0$ hence

$$(D_1^2 + D_2^2)\varphi_\epsilon(x_1, y_1) = 2\epsilon > 0.$$

This contradiction ends the proof.

Exercise VIII.1.6. *Let*

$$u(x, y) = \operatorname{Re} \frac{i+z}{i-z} \quad \text{for } z \neq i \text{ and } u(0, 1) = 0.$$

Show that u is harmonic on the unit disc, is 0 on the unit circle, and is continuous on the closed unit disc except at the point $z = i$. This gives a counterexample when u is not bounded.

Solution. Since $f : z \mapsto (i + z)/(i - z)$ is analytic on D, the function u is harmonic on D. Let $z = x + iy$. Then

$$f(z) = \frac{x + i(y+1)}{-x + i(1-y)} = \frac{[x + i(y+1)][-x - i(1-y)]}{x^2 + (1-y)^2},$$

and therefore we get the formula

$$u(x, y) = \frac{-x^2 + 1 - y^2}{x^2 + (1-y)^2}.$$

If (x, y) belongs to the unit circle then $x^2 + y^2 = 1$, and if $(x, y) \neq (0, 1)$, we see from the above expression that $u(x, y) = 0$. Moreover we assume $u(0, 1) = 0$ so u is identically zero on the unit circle.

From the above formula for u we see that this function is continuous on the closed unit disc except at i. Indeed, this is the only point that cancels the denominator, and if $x = 0$ we see that

$$\frac{1 - y^2}{(1-y)^2} = \frac{1+y}{1-y} \to \infty \quad \text{as } y \to 1.$$

Exercise VIII.1.7. *Find an analytic function whose real part is the given function.*
(a) $u(x, y) = 3x^2y - y^3$
(b) $x - xy$
(c) $\frac{y}{x^2+y^2}$
(d) $\log \sqrt{x^2 + y^2}$
(e) $\frac{y}{(x-t)^2+y^2}$ where t is some real number.

Solution. (a) Let $f = u + iv$ where $v(x, y) = 3xy^2 - x^3$. Then the Cauchy–Riemann equations imply that f is analytic.

(b) Let $f = u + iv$ where $v(x, y) = y - y^2/2 + x^2/2$. Then the Cauchy–Riemann equations imply that f is analytic.

(c) Let $f(z) = i/z$. Then

$$f(z) = \frac{i}{x + iy} = \frac{y}{x^2 + y^2} + i \frac{x}{x^2 + y^2}.$$

(d) The choice $f(z) = \log z$ is a solution.

(e) The choice $f(z) = \frac{i}{z-t}$ will do.

Exercise VIII.1.8. *Let $f(z) = \log z$. If $z = re^{i\theta}$, then*

$$f(z) = \log r + i\theta,$$

so the real parts and imaginary parts are given by

$$u = \log r \quad \text{and} \quad v = \theta.$$

Draw the level curves $u =$ constant and $v =$ constant. Observe that they intersect orthogonally.

Solution. The level curves are

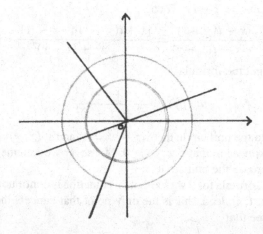

The circles correspond to the level curves of u, and the half lines are the level curves of v.

Exercise VIII.1.9. *Let V be the open set obtained by deleting the segment $[0, 1]$ from the right half plane, as shown on the figure. In other words V consists of all complex numbers $x + iy$ with $x > 0$, with the exception of the numbers $0 < x \leq 1$.*

(a) What is the image of V under the map $z \mapsto z^2$?

(b) What is the image of V under the map $z \mapsto z^2 - 1$?

(c) Find an isomorphism of V with the right half plane, and then with the upper half plane. [Hint: Consider the function $z \mapsto \sqrt{z^2 - 1}$.]

Solution. (a) If $z = re^{i\theta}$, then $z^2 = r^2 e^{2i\theta}$ so the image of V under the map $z \mapsto z^2$ is the complex plane minus the half line $(-\infty, 1]$.

(b) From (a) we see that the image of V under the map $z \mapsto z^2 - 1$ is $\mathbf{C} - \mathbf{R}_{\leq 0}$.

(c) Since $\mathbf{C} - \mathbf{R}_{\leq 0}$ is simply connected we can define $z \mapsto \sqrt{z}$ and if $z = re^{i\theta}$ with $r > 0$ and $-\pi < \theta < \pi$, then $\sqrt{z} = \sqrt{r}e^{i\theta/2}$ so the image of $\mathbf{C} - \mathbf{R}_{\leq 0}$ is the right half plane. Since the map $i \mapsto iz$ is a rotation by an angle of $\pi/2$ around the origin, the image of V under the map $z \mapsto i\sqrt{z^2 - 1}$ is the upper half plane.

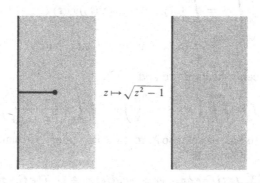

Exercise VIII.1.10. *Let U be the open set discussed at the end of the section, obtained by deleting the vertical segment of points $(0, y)$ with $0 \leq y \leq 1$ from the upper half plane. Find an analytic isomorphism*

$$f : U \to H.$$

[Hint: Rotate the picture $90°$ and use Exercise 9.]

Solution. With the notation of Exercise 9 we have $V = -iU$ so the map

$$f : z \mapsto i\sqrt{(-iz)^2 - 1}$$

is an analytic isomorphism $U \to H$.

Exercise VIII.1.11. *Let φ be a complex harmonic function on a connected open set U. Suppose that φ^2 is also harmonic. Show that φ or $\overline{\varphi}$ is holomorphic.*

Solution. Copy the argument given in Exercise VIII.1.1 (b) by replacing $\mathbf{C}$ by U. Since U is connected, the argument caries over.

Exercise VIII.1.12. *Green's Theorem in calculus states: Let $p = p(x, y)$ and $q = q(x, y)$ be C^1 functions on the closure of a bounded open set U whose boundary consists of a finite number of C^1 curves oriented so that U lies to the left of each one of these curves. Let C be this boundary. Then*

$$\int_C p\,dx + q\,dy = \iint_U \left(\frac{\partial q}{\partial x} - \frac{\partial p}{\partial y} \right) dy\,dx.$$

Suppose that f is analytic on U and on its boundary. Show that Green's theorem implies Cauchy's theorem for the boundary, i.e., show that

$$\int_C f = 0.$$

Solution. Let u and v be the real and imaginary parts of f respectively. Let $\gamma : [a, b] \to \mathbf{C}, t \mapsto \gamma_1(t) + i\gamma_2(t)$ be a parametrization of C. Then by definition

$$\int_C f = \int_a^b (u(\gamma(t)) + iv(\gamma(t)))\gamma'(t)dt$$

$$= \int_C (udx - vdy) + i(vdx + udy).$$

Now applying Green's theorem, we find

$$\int_C f = \int\int_U \left(-\frac{\partial v}{\partial x} - \frac{\partial u}{\partial y}\right) + i \int\int_U \left(\frac{\partial u}{\partial x} - \frac{\partial v}{\partial y}\right).$$

But the function f is holomorphic so its real and imaginary part satisfy the Cauchy–Riemann equations. Conclude.

Exercise VIII.1.13. *Let U be an open set and let $z_0 \in U$. The **Green's function for U originating at** z_0 is a real function g defined on the closure $\overline{U}$ of U, continuous except at z_0, and satisfying the following conditions:*

GR 1. *$g(z) = \log|z - z_0| + \psi(z)$, where ψ is harmonic on U.*

GR 2. *g vanishes on the boundary of U.*

(a) Prove that a Green's function is uniquely determined if U is bounded.
(b) Let U be simply connected, with smooth boundary. Let

$$f : U \to D$$

be an analytic isomorphism of U with the unit disc such that $f(z_0) = 0$. Let

$$g(z) = \operatorname{Re} \log f(z).$$

Show that g is a Green's function for U. You may assume that f extends to a continuous function from the boundary of U to the boundary of D.

Solution. (a) Suppose U is bounded and g_1 and g_2 are two Green's functions for U originating at z_0. Write

$$g_i(z) = \log|z - z_0| + \psi_i(z) \quad \text{for } i = 1, 2$$

where ψ_i is harmonic on U and $g_i(z) = 0$ on the boundary of U. Let $h(z) = g_1(z) - g_2(z)$. Then $h(z) = \psi_1(z) - \psi_2(z)$ so h is harmonic on U and $h = 0$ on the boundary of U. Since U is bounded, Theorem 1.3 implies that $h = 0$, so $g_1 = g_2$ as was to be shown.
(b) Since f is an analytic isomorphism which extends from the boundary of U to the boundary of D we see that if $z \in \partial U$, then $|f(z)| = 1$ hence $g(z) = 0$, proving

that **GR 1** holds. Also, since f is an analytic isomorphism, z_0 is the unique solution of the equation $f(z) = 0$ in U, and $f'(z) \neq 0$ for all $z \in U$. Define a function α by

$$\alpha(z) = \frac{f(z)}{z - z_0}.$$

The function α is analytic on U and has no zeros. Therefore the function $\log \alpha(z)$ is well defined and analytic on U which is simply connected. We have

$$g(z) = \operatorname{Re} \log f(z) = \log |f(z)| = \log |z - z_0| + \log |\alpha(z)|$$
$$= \log |z - z_0| + \operatorname{Re} \log \alpha(z),$$

but $\operatorname{Re} \log \alpha(z)$ is harmonic on U so **GR 2** holds.

VIII.2   Examples

Exercise VIII.2.1. *Find a harmonic function of the upper half plane with value 1 on the positive real axis and value -1 on the negative real axis.*

Solution. We can take $\varphi(z) = 1 - \frac{2}{\pi} \arg z$.

Exercise VIII.2.2. *Find a harmonic function on the indicated region, with the boundary values as shown.*

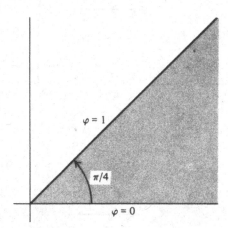

Solution. The map $z \mapsto z^4$ reduces the problem to the case where we have the upper half plane, with $\varphi = 0$ on $\mathbf{R}_{>0}$ and $\varphi = 1$ on $\mathbf{R}_{<0}$, so we can take $\frac{1}{\pi} \arg z$. Therefore a solution to our problem is $\varphi(z) = \frac{1}{\pi} \arg z^4$.

Exercise VIII.2.3. *Find the temperature on a semicircular plate of radius 1, as shown on the figure, with the boundary values as shown. Value 0 on the semicircle, value 1 on one segment, value 0 on the other segment.*

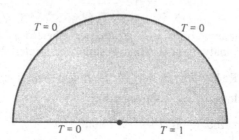

Solution. There exists an isomorphism f mapping the upper half unit disc onto the upper half plane such that $(0, 1) \to \mathbf{R}_{>0}$ and also upper half circle $\cup[-1, 0) \to \mathbf{R}_{<0}$. Indeed, f is obtained by using Example 3, §4 of Chapter VII (see also Exercise VII.3.1) together with $z \mapsto z^2$ and a translations. We have the following sequence

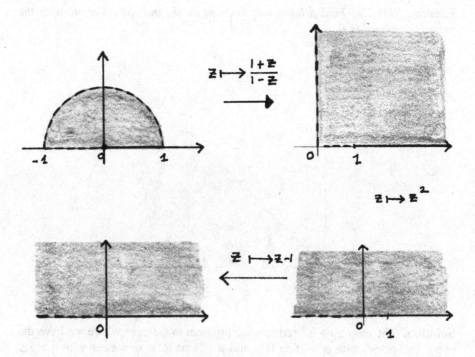

In the upper half plane we take $\frac{1}{\pi}\arg z$, so an answer to this exercise is $\frac{1}{\pi}(\pi - \arg f(z))$.

Exercise VIII.2.4. *Find a harmonic function on the unit disc which has the boundary value 0 on the lower semicircle and boundary value 1 on the upper semicircle.*

Solution. One verifies easily that $f : z \mapsto -i(z+1)/(z-1)$ is an isomorphism of the unit disc with the upper half plane. If $z = x + iy$ then a direct computation shows that

$$f(z) = \frac{-2y}{(x-1)^2 + y^2} + i\frac{1 - x^2 + y^2}{(x-1)^2 + y^2}.$$

Thus if z belongs to the unit circle we have

$$f(z) = \frac{-y}{1-x}.$$

So we are reduced to the case where we have the upper half plane with $\varphi = 0$ on $\mathbf{R}_{>0}$ and $\varphi = 1$ on $\mathbf{R}_{<0}$. We can choose $\frac{1}{\pi}\arg z$, so an answer to the exercise is $\frac{1}{\pi}\arg f(z)$.

In the next exercise, recall that a function $\varphi : U \to \mathbf{R}$ is said to be of class C^1 if its partial derivatives $D_1\varphi$ and $D_2\varphi$ exist and are continuous. Let V be another open set. A mapping

$$f : V \to \mathbf{R}^2$$

where $f(x, y) = (u(x, y), v(x, y))$ is said to be of class C^1 if the two coordinate functions u, v are of class C^1.

If $\eta : [a, b] \to V$ is a curve in V, the we may form the composite curve $f \circ \eta$ such that

$$(f \circ \eta)(t) = f(\eta(t)).$$

Then $\gamma = f \circ \eta$ is a curve in U. Its coordinates are

$$f(\eta(t)) = (u(\eta(t)), v(\eta(t))).$$

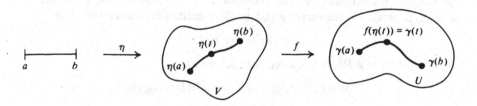

Exercise VIII.2.5. *Let $\gamma : [a, b] \to \mathbf{R}^2$ be a smooth curve. Let*

$$\gamma(t) = (\gamma_1(t), \gamma_2(t))$$

*be the expression of γ in terms of its coordinates. The **tangent vector** is given by the derivative $\gamma'(t) = (\gamma_1'(t), \gamma_2'(t))$. We define*

$$N(t) = (\gamma_2'(t), -\gamma_1'(t))$$

*to be the **normal vector**. We define the **unit normal vector** to be*

$$n(t) = \frac{N(t)}{|N(t)|}, \quad \text{where } |N(t)| = \sqrt{N_1(t)^2 + N_2(t)^2},$$

assuming throughout that $|\gamma'(t)| \neq 0$ for all t. Verify that $\gamma'(t) \cdot N(t) = 0$.

If γ is a curve in an open set U, and φ is of class C^1 on U, we define

$$D_n\varphi = \frac{\partial \varphi}{\partial n} = (\operatorname{grad} \varphi) \cdot n$$

*to be the right hand **normal derivative** of φ along the curve.*

(a) Prove that if $\partial \varphi/\partial n = 0$, then this condition remains true under a change of parametrization.

(b) Let $f : V \to U$ be analytic. Let η be a curve in V and let $\gamma = f \circ \eta$. If $D_n\varphi = 0$ on γ, show that $D_n(\varphi \circ f) = 0$ on η. [Hint: Prove a stronger result. Fix some value $t_0 \in [a, b]$. Show first that $(D_n\varphi)(\gamma(t_0)) = 0$ if and only if there is a real number c such that $(\operatorname{grad} \varphi)(\gamma(t_0)) = c\gamma'(t_0)$. In other words, for all $w \in \mathbf{C} = \mathbf{R}^2$, we have $\varphi'(\gamma(t_0))w = \langle c\gamma'(t_0), w \rangle$. Cf. Chapter I, §7 (1). Then use the chain rule to compute $(\varphi \circ f)'(\eta(t_0))x$ with arbitrary $z \in \mathbf{C} = \mathbf{R}^2$.]

Solution. (a) Let $\beta = \gamma \circ g$ be a reparametrization of the curve γ. Then

$$(\operatorname{grad} \varphi) \cdot N_\beta = \frac{\partial \varphi}{\partial x}(\beta(t))\beta_2'(t) - \frac{\partial \varphi}{\partial y}(\beta(t))\beta_1'(t),$$

but by the chain rule we have $\beta_i' = \gamma_i'(g(t))g'(t)$ so the above expression is equal to

$$g'(t)\left(\frac{\partial \varphi}{\partial x}(\gamma(g(t)))\gamma_1'(g(t)) - \frac{\partial \varphi}{\partial y}(\gamma(g(t)))\gamma_2'(g(t))\right).$$

The expression in the big parenthesis is 0 by assumption so $(\operatorname{grad} \varphi) \cdot N_\beta = 0$ and therefore the condition $\partial \varphi/\partial n = 0$ remains true under a change of parametrization.

(b) Fix t_0 in the domain of γ. The condition $D_n\varphi = 0$ says that the gradient of φ at $\gamma(t_0)$ is parallel to the vector $\gamma'(t_0)$, so there exists a constant c such that

$$\varphi'(t_0)w = c\langle \gamma'(t_0), w \rangle$$

for all vectors $w \in \mathbf{R}^2$. If z is a vector in $\mathbf{R}^2$ we have

$$(\varphi \circ f)'(\eta(t_0))z = \varphi'(f(\eta(t_0)))Df(\eta(t_0))z,$$

but f is holomorphic, so

$$Df(\eta(t_0))z = f'(\eta(t_0))z$$

where the multiplication on the left is that of matrices and the multiplication on the right is that of complex numbers and where the identification of $\mathbf{R}^2$ and $\mathbf{C}$ is the usual one. From this, we obtain

$$
\begin{aligned}
(\varphi \circ f)'(\eta(t_0))z &= \varphi'(f(\eta(t_0)))f'(\eta(t_0))z \\
&= \varphi'(\gamma(t_0))f'(\eta(t_0))z \\
&= c\langle \gamma'(t_0), f'(\eta(t_0))z \rangle \\
&= c\langle f'(\eta(t_0))\eta'(t_0), f'(\eta(t_0))z \rangle \\
&= c|f'(\eta(t_0))|^2 \langle \eta'(t_0), z \rangle,
\end{aligned}
$$

where for complex numbers z_1, z_2 we define $\langle z_1, z_2 \rangle = \mathrm{Re}(z_1\bar{z}_2)$ (this corresponds to the usual scalar product in $\mathbf{R}^2$).

This shows that the gradient of $\varphi \circ f$ at $\eta(t_0)$ is parallel to $\eta'(t_0)$ and we conclude that $\partial(\varphi \circ f)/\partial n = 0$ on η. Note that this computation also describes the change of the gradient under the mapping f, namely the factor $|f'(\eta(t_0))|^2$ which corresponds to the determinant of the Jacobian of f.

One could also argue as follows. Let A be the 2×2 matrix defined by

$$
A = \begin{pmatrix} 0 & -1 \\ 1 & 0 \end{pmatrix}.
$$

Then note that $A\gamma' = N$, so the condition $\partial\varphi/\partial n = 0$ can be written as

$$
0 = \varphi' A\gamma' = \varphi' A(Df)\eta'.
$$

Similarly

$$
\begin{aligned}
\frac{\partial(\varphi \circ f)}{\partial n} &= (\varphi \circ f)' A\eta' \\
&= \varphi'(Df)A\eta'.
\end{aligned}
$$

But f is holomorphic, so its real and imaginary parts satisfy the Cauchy–Riemann equations and this means $A(Df) = (Df)A$. We conclude that $\partial(\varphi \circ f)/\partial n = 0$.

Exercise VIII.2.6. *Find a harmonic function φ on the indicated regions, with the indicated boundary values. (Recall what $\sin z$ does to the vertical strip.) See the figures on pages 158–159.*

Solution. (a) It suffices to find an isomorphism which takes the given region to the upper half plane, and where the piece of the boundary where $\varphi = 0$ corresponds to the positive reals, and the piece of the boundary where $\varphi = 1$ corresponds to the negative reals. Then we may take $u(z) = (1/\pi)\arg(z)$. We now show how to obtain this isomorphism.

First, the isomorphism $w \mapsto (w-1)/(w+1)$ maps the first quadrant to the upper semidisc. Then apply $z \mapsto \log z$ to map this semidisc to a half strip. Now rotate this vertical strip, and translate it so that it becomes the vertical strip $\{x + iy : -\pi/2 < x < \pi/2, \ 0 < y\}$. Now apply $z \mapsto \sin z$ to this strip and translate to obtain the upper half plane with the desired boundary values. We keep track of the boundary values on the pictures on page 160.

(a)

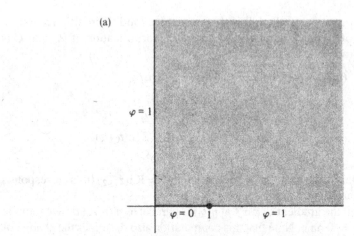

$\varphi = 1$

$\varphi = 0$ 1 $\varphi = 1$

(b)

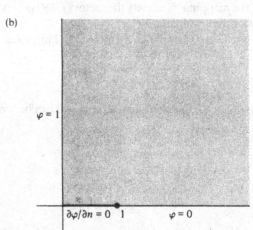

$\varphi = 1$

$\partial\varphi/\partial n = 0$ 1 $\varphi = 0$

(c)

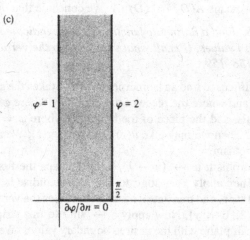

$\varphi = 1$ $\varphi = 2$

$\dfrac{\pi}{2}$

$\partial\varphi/\partial n = 0$

(d)

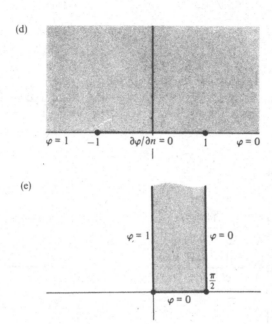

$\varphi = 1$ -1 $\partial \varphi / \partial n = 0$ 1 $\varphi = 0$

(e)

$\varphi = 1$ $\varphi = 0$

$\frac{\pi}{2}$

$\varphi = 0$

(b) It suffices to map the given region to a half vertical strip with boundary value 1 on the left vertical segment, $\varphi = 0$ on the right vertical segment and $\partial \varphi / \partial n = 0$ on the horizontal segment. Then we may take the function $1 - x$ as an answer. The sequence of pictures on pages 158–159 describes the isomorphism. When there is no mention of the map, we simply rotate, dilate and translate.
(c) The function $(2/\pi)x + 1$ will work.
(d) This domain appears in part (b) as we can see from the pictures. Pick up the sequence of isomorphisms from there, and take at the end the function $1 - x$.
(e) Here, use a translation, a dilation and $\sin z$ with another translation to end up with the upper half plane with $\varphi = 0$ on the positive reals and $\varphi = 1$ on the negative reals. Set $u(x) = (1/\pi) \arg z$.

VIII.3   Basic Properties of Harmonic Functions

Exercise VIII.3.1. *The **Gauss theorem** (a variation of Green's theorem) can be stated as follows.*
Let γ be a closed piecewise C^1 curve in an open set U, and suppose γ has an interior contained in U. Let F be a C^1 vector field on U. Let n be the unit normal

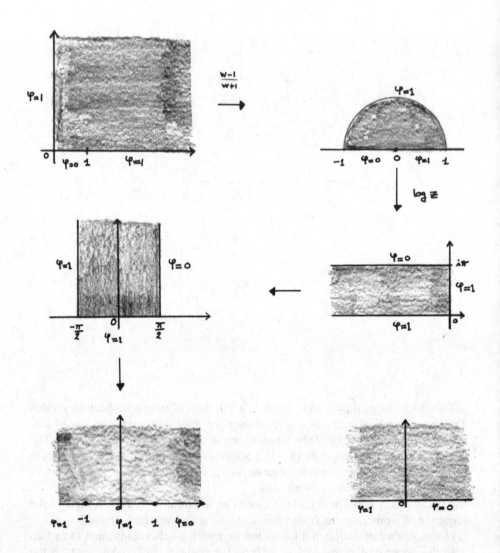

vector on γ. *Then*

$$\int_\gamma F \cdot n = \iint_{\mathrm{Int}(\gamma)} (\mathrm{div}\ F) dy dx.$$

Using the Gauss theorem, prove the following. Let u be a C^2 *function on U, harmonic on the interior* $\mathrm{Int}(\gamma)$. *Then*

$$\int_\gamma D_n u = 0.$$

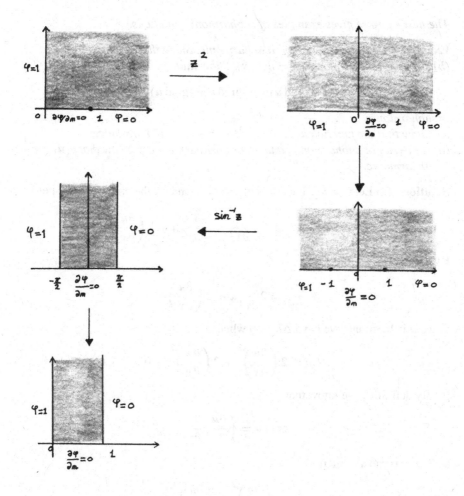

Here $D_n u$ is the normal derivative $(\text{grad } u) \cdot n$, as in Exercise 5 of §2.

Solution. Apply Gauss's theorem to the vector field $F = \text{grad } u$ and note that

$$\text{div } F = \frac{\partial^2}{\partial x^2} u + \frac{\partial^2}{\partial y^2} u = 0.$$

Subharmonic Functions

*Define a real function φ to be **subharmonic** if φ is of class C^2 (i.e., has continuous partial derivatives up to order 2) and*

$$\frac{\partial^2 \varphi}{\partial x^2} + \frac{\partial^2 \varphi}{\partial y^2} \geq 0.$$

The next exercise gives examples of subharmonic functions.

Exercise VIII.3.2. *(a) Let u be real harmonic. Show that u^2 is subharmonic.*
(b) Let u be real harmonic, $u = u(x, y)$. Show that

$$(\operatorname{grad} u)^2 = (\operatorname{grad} u) \cdot (\operatorname{grad} u)$$

is subharmonic.
(c) Show that the function $u(x, y) = x^2 + y^2 - 1$ is subharmonic.
(d) Let u_1, u_2 be subharmonic, and c_1, c_2 positive numbers. Show that $c_1 u_1 + c_2 u_2$ is subharmonic.

Solution. (a) Let $v = u^2$. Then v is of class C^2 and by the chain rule we get

$$\frac{\partial v}{\partial x} = 2 \frac{\partial u}{\partial x} u \quad \text{and} \quad \frac{\partial^2 v}{\partial x^2} = 2 \frac{\partial^2 u}{\partial x^2} u + 2 \left(\frac{\partial u}{\partial x} \right)^2.$$

By symmetry we find

$$\frac{\partial^2 v}{\partial y^2} = 2 \frac{\partial^2 u}{\partial y^2} u + 2 \left(\frac{\partial u}{\partial y} \right)^2.$$

Since u is harmonic we have $\Delta u = 0$ whence

$$\Delta v = 2 \left(\frac{\partial u}{\partial x} \right)^2 + 2 \left(\frac{\partial u}{\partial y} \right)^2 \geq 0.$$

(b) By definition we know that

$$\operatorname{grad} u = \left(\frac{\partial u}{\partial x}, \frac{\partial u}{\partial y} \right)$$

so if $v = (\operatorname{grad} u)^2$, then

$$v = \left(\frac{\partial u}{\partial x} \right)^2 + \left(\frac{\partial u}{\partial y} \right)^2.$$

Applying the chain rule we find

$$\frac{\partial^2 v}{\partial x^2} = 2 \left[\frac{\partial^3 u}{\partial x^3} \frac{\partial u}{\partial x} + \left(\frac{\partial^2 u}{\partial x^2} \right)^2 + \frac{\partial^3 u}{\partial y^2 \partial x} \frac{\partial u}{\partial x} + \left(\frac{\partial^2 u}{\partial x \partial y} \right)^2 \right].$$

We get a symmetric expression for $\partial^2 v / \partial y^2$. Adding both terms, and using the fact that $\Delta u =$ we find that $\Delta v \geq 0$ as was to be shown.
(c) We simply have $\Delta u = 4 \geq 0$.
(d) Differentiation is a linear operator so we get

$$\Delta(c_1 u_1 + c_2 u_2) = c_1 \Delta u_1 + c_2 \Delta u_2.$$

The numbers c_1 and c_2 are positive, and u_1 and u_2 are subharmonic so $\Delta(c_1 u_1 + c_2 u_2) \geq 0$.

Exercise VIII.3.3. *Let φ be subharmonic on an open set containing a closed disc of radius r_1 centered at a point a. For $r < r_1$ let*

$$h(r) = \int_0^{2\pi} \varphi(a + re^{i\theta}) \frac{d\theta}{2\pi}.$$

Show that $h(r)$ is increasing as a function of r. [Hint: Let $u(r, \theta) = \varphi(a + re^{i\theta})$. Then

$$r \frac{d}{dr}(rh'(r)) = \int_0^{2\pi} r \frac{\partial}{\partial r} \left(r \frac{\partial u}{\partial r} \right) \frac{d\theta}{2\pi}.$$

Use the expression for Δ in polar coordinates, and the fact that the integral of $\partial^2 u/\partial\theta^2$ is 0 to show that $rh'(r)$ is weakly increasing. Since $rh'(r) = 0$ for $r = 0$, it follows that $rh'(r) \geq 0$, so $h'(r) \geq 0$.]

Solution. The Laplacian in polar coordinates is given by

$$\Delta = \frac{\partial^2}{\partial r^2} + \frac{1}{r} \frac{\partial}{\partial r} + \frac{1}{r^2} \frac{\partial^2}{\partial \theta^2}.$$

Now

$$r \frac{\partial}{\partial r} \left(r \frac{\partial u}{\partial r} \right) = r \frac{\partial u}{\partial r} + r^2 \frac{\partial^2 u}{\partial r^2},$$

and since $r^2 \Delta u \geq 0$ we conclude that

$$r \frac{d}{dr}(rh'(r)) \geq \int_0^{2\pi} r \frac{\partial u}{\partial r} - r \frac{\partial u}{\partial r} - \frac{\partial^2 u}{\partial \theta^2} \frac{d\theta}{2\pi} = 0.$$

The fact that the integral of $\partial^2 u/\partial\theta^2$ is 0 follows from the fundamental theorem of calculus and the fact that u is periodic.

Exercise VIII.3.4. *Using Exercise 3, or any other way, prove the inequality*

$$\varphi(a) \leq \int_0^{2\pi} \varphi(a + re^{i\theta}) \frac{d\theta}{2\pi} \quad \text{for every } r.$$

Solution. With the notation of Exercise 3 we have $h(0) \leq h(r)$ because h is differentiable and $h'(r) \geq 0$. Moreover, $h(0) = \varphi(a)$, so we get the desired inequality.

Exercise VIII.3.5. *Suppose that φ is defined on an open set U and is subharmonic on U. Prove the maximum principle, that no point $a \in U$ can be a strict maximum for φ, i.e., that for every disc of radius r centered at a with r sufficiently small, we have*

$$\varphi(a) \leq \max \varphi(z) \quad \text{for } |z - a| = r.$$

Solution. Let $a \in U$, and let r be so small that the closed disc $\overline{D}_r(a)$ centered at a and of radius r is contained in U. Since φ is continuous on the compact set $C_r(a)$, the boundary of $\overline{D}_r(a)$, it attains its maximum on that set and therefore

$$\varphi(a + re^{i\theta}) \leq \max_{z \in C_r(a)} \varphi(z)$$

for all θ. Integrating with respect to θ from 0 to 2π and using the result of Exercise 4 we find

$$\varphi(a) \leq \frac{1}{2\pi} \int_0^{2\pi} \varphi(a + re^{i\theta})d\theta \leq \max_{z \in C_r(a)} \varphi(z).$$

Conclude.

Exercise VIII.3.6. *Let φ be subharmonic on an open set U. Assume that the closure $\overline{U}$ is compact, and that φ extends to a continuous function on $\overline{U}$. Show that the maximum for φ occurs on the boundary.*

Solution. The function φ is continuous on the compact set $\overline{U}$ and therefore attains its maximum on $\overline{U}$. Suppose that φ attains its maximum value at an interior point a of U. We claim that there exists $z \in \partial U$ such that $\varphi(z) = \varphi(a)$.

First we show that φ is locally constant at a. Select r so small that the closed disc $\overline{D}_r(a)$ is contained in U. Consider the function f defined on the boundary of $\overline{D}_r(a)$ by

$$f(\theta) = \varphi(a) - \varphi(a + re^{i\theta}).$$

By assumption $f \geq 0$. Suppose that there exists $0 \leq \theta_0 \leq 2\pi$ such that $f(\theta_0) > 0$. Then by continuity we have

$$\int_0^{2\pi} f(\theta)d\theta > 0.$$

This inequality combined with the result obtained in Exercise 4 we find

$$\varphi(a) \leq \frac{1}{2\pi} \int_0^{2\pi} \varphi(a + re^{i\theta})d\theta < \varphi(a),$$

which is a contradiction and therefore $f = 0$. This proves that φ is locally constant.

Now let V be the largest connected open set in U containing a. By the method of propagation along curves and the fact that φ is locally constant, we see that $\varphi(z) = \varphi(a)$ for all $z \in V$. Since $\overline{V} \subset \overline{U}$ the boundary of V, ∂V is nonempty and we contend that $\partial V \subset \partial U$. If not, then we can find $z \in \partial V$ with $z \in U$. Taking a small ball $B \subset U$ containing z we see that $V \cup B$ is open, connected and contained in U. This contradicts the maximality of V and proves our contention. By continuity, we have $\varphi = \varphi(a)$ on ∂V and therefore there exists $z \in \partial U$ such that $\varphi(z) = \varphi(a)$ as was to be shown.

Exercise VIII.3.7. *Let U be a bounded open set. Let u, v be continuous functions on $\overline{U}$ such that u is harmonic on U, v is subharmonic on U, and $u = v$ on the boundary of U. Show that $v \leq u$ on U. Thus a subharmonic functions lies below the harmonic function having the same boundary value, whence its name.*

Solution. The function $v - u$ is subharmonic because

$$\Delta(v - u) = \Delta v - \Delta u = \Delta v \geq 0.$$

On the boundary of U we have $v - u \leq 0$, so by Exercise 6 it follows that $v - u \leq 0$ on U as was to be shown.

VIII.4   The Poisson Formula

Exercise VIII.4.1. *Give another proof of Theorem 4.1 as follows. First by Cauchy's theorem,*

$$f(0) = \frac{1}{2\pi i} \int_{C_R} \frac{f(\zeta)}{\zeta} d\zeta.$$

Let g be the automorphism of the disc which interchanges 0 and z. Apply the above formula to the function f ∘ g instead of f, and change variables in the integral, with $w = g(\zeta)$, $\zeta = g^{-1}(w)$.

Solution. Using the map $D_R \to D$ defined by $z \mapsto z/R$ we see that the automorphism of D_R interchanging 0 and z is given by

$$g_z(w) = R\frac{\frac{z}{R} - \frac{w}{R}}{1 - \frac{\bar{z}}{R}\frac{w}{R}} = \frac{z - w}{1 - \frac{\bar{z}w}{R^2}}.$$

By Cauchy's formula,

$$f(z) = f(g_z(0)) = \frac{1}{2\pi i} \int_{C_R} \frac{f(g_z(\zeta))}{\zeta} d\zeta.$$

Since $g_z = g_z^{-1}$ we can make the change of variable $\zeta = g_z(w)$ in the integral and we get

$$f(z) = \frac{1}{2\pi i} \int_{C_R} \frac{f(w)}{g_z(w)} g_z'(w) dw.$$

A direct computation shows that

$$\frac{g_z'(w)}{g_z(w)} = \frac{-1 + \frac{\bar{z}z}{R^2}}{\left(1 - \frac{\bar{z}w}{R^2}\right)^2} \cdot \frac{1 - \frac{\bar{z}w}{R^2}}{z - w}$$

$$= \frac{-1 + \frac{\bar{z}z}{R^2}}{\left(1 - \frac{\bar{z}w}{R^2}\right)(z - w)}$$

and therefore letting $w = Re^{i\theta}$ in the integral we get

$$f(z) = \frac{1}{2\pi} \int_0^{2\pi} f(Re^{i\theta}) \frac{-1 + \frac{\bar{z}z}{R^2}}{\left(1 - \frac{\bar{z}w}{R^2}\right)(z - w)} w d\theta,$$

so it is sufficient to show that

$$\text{Re}\left(\frac{w + z}{w - z}\right) = \frac{-1 + \frac{\bar{z}z}{R^2}}{\left(1 - \frac{\bar{z}w}{R^2}\right)(z - w)} w.$$

The left hand side is equal to

$$\text{Re}\left(\frac{w + z}{w - z}\right) = \frac{1}{2}\left(\frac{w + z}{w - z} + \frac{\bar{w} + \bar{z}}{\bar{w} - \bar{z}}\right) = \frac{w\bar{w} - z\bar{z}}{(w - z)(\bar{w} - \bar{z})},$$

and since $R^2 = w\overline{w}$, the right hand side is equal to

$$\frac{-1 + \frac{\overline{z}z}{R^2}}{\left(1 - \frac{\overline{z}w}{R^2}\right)(z - w)}w = \frac{-1 + \frac{\overline{z}z}{w\overline{w}}}{\left(1 - \frac{\overline{z}w}{Rw\overline{w}}\right)(z - w)}w = \frac{w\overline{w} - z\overline{z}}{(w - z)(\overline{w} - \overline{z})}.$$

This concludes the exercise.

Exercise VIII.4.2. *Define*

$$P_{R,r}(\theta) = \frac{1}{2\pi}\frac{R^2 - r^2}{R^2 - 2Rr\cos\theta + r^2}$$

for $0 \le r < R$.

Prove the inequalities

$$\frac{R - r}{R + r} \le 2\pi P_{R,r}(\theta - \varphi) \le \frac{R + r}{R - r}$$

for $0 \le r < R$.

Solution. We have

$$-2rR \le -2rR\cos(\theta - \varphi) \le 2rR$$

hence

$$(R - r)^2 \le R^2 - 2rR\cos(\theta - \varphi) + r^2 \le (R + r)^2.$$

The inequalities $0 \le r \le R$ imply

$$\frac{R^2 - r^2}{(R + r)^2} \le 2\pi P_{R,r}(\theta - \varphi) \le \frac{R^2 - r^2}{(R - r)^2},$$

and since $R^2 - r^2 = (R + r)(R - r)$ we get the desired inequalities.

Exercise VIII.4.3. *Let f be analytic on the closed disc $\overline{D}(\alpha, R)$ and let $u = \mathrm{Re}(f)$. Assume that $u \ge 0$. Show that for $0 \le r < R$ we have*

$$\frac{R - r}{R + r}u(\alpha) \le u(\alpha + re^{i\theta}) \le \frac{R + r}{R - r}u(\alpha).$$

After you have read the next section, you will see that this inequality holds also if $u \ge 0$ is harmonic on the disc, with a continuous extension to the closed disc $\overline{D}(\alpha, R)$.

Solution. After translation we may assume that $\alpha = 0$. The function u is harmonic because it is the real part of an analytic function so by Theorem 4.2 we have with the notation of Exercise 2

$$u(re^{i\varphi}) = \int_0^{2\pi} u(re^{i\theta})P_r(\theta - \varphi)d\theta.$$

The inequalities of Exercise 2 combined with the fact that $u \ge 0$ give

$$\int_0^{2\pi} u(re^{i\theta})\frac{R - r}{R + r}\frac{d\theta}{2\pi} \le u(re^{i\varphi}) \le \int_0^{2\pi} u(re^{i\theta})\frac{R + r}{R - r}\frac{d\theta}{2\pi}.$$

The mean value theorem for harmonic functions states that

$$u(0) = \int_0^{2\pi} u(re^{i\theta})\frac{d\theta}{2\pi}.$$

Conclude.

Exercise VIII.4.4. *Let $\{u_n\}$ be a sequence of harmonic functions on the open disc. If it converges uniformly on compact subsets of the disc, then the limit is harmonic.*

Solution. Let $0 < r_0 < r_1 < 1$ and choose functions f_n, analytic on the unit disc, and such that $\operatorname{Re}(f_n) = u_n$. After subtracting an imaginary constant if necessary, we may assume (by Theorem 4.2) that for $z \in D_{r_0}$

$$f_n(z) = \int_0^{2\pi} u_n(r_1e^{i\theta})\frac{r_1e^{i\theta} + z}{r_1e^{i\theta} - z}\frac{d\theta}{2\pi}.$$

For $z \in D_{r_0}$ let

$$f(z) = \int_0^{2\pi} u(r_1e^{i\theta})\frac{r_1e^{i\theta} + z}{r_1e^{i\theta} - z}\frac{d\theta}{2\pi},$$

where u is the limit of $\{u_n\}$. Then, for $z \in D_{r_0}$ we have

$$|f_n(z) - f(z)| \leq \int_0^{2\pi} |u_n(r_1e^{i\theta}) - u(r_1e^{i\theta})| \left|\frac{r_1e^{i\theta} + z}{r_1e^{i\theta} - z}\right| \frac{d\theta}{2\pi}$$

$$\leq \|u_n - u\|_{\overline{D}_{r_1}} \frac{2}{r_1 - r_0}$$

where $\| \cdot \|_{\overline{D}_{r_1}}$ denotes the sup norm on $\overline{D}_{r_1}$. So $f_n \to f$ uniformly on compact subsets of D_{r_0} whence f is holomorphic on D_{r_0} and $\operatorname{Re}(f) = u$ because

$$\operatorname{Re}(f) = \lim \operatorname{Re}(f_n) = \lim u_n = u.$$

We conclude that u is harmonic on D_{r_0}, and since r_0 was chosen arbitrarily we get that u is harmonic on the unit disc.

VIII.5 Construction of Harmonic Functions

One can also consider Dirac sequences or families over the whole real line. We use a notation which will fit a specific application. For each $y > 0$ suppose given a continuous function P_y on the real line, satisfying the following conditions:

DIR 1. $P_y(t) \geq 0$ *for all y, and all real t.*

DIR 2. $\int_{-\infty}^{\infty} P_y(t)dt = 1.$

DIR 3. *Given ϵ, δ there exists $y_0 > 0$ such that if $0 < y < y_0$, then*

$$\int_{-\infty}^{-\delta} + \int_{\delta}^{\infty} P_y(t)dt < \epsilon.$$

*We call $\{P_y\}$ a **Dirac family** again, for $y \to 0$. Prove:*

Exercise VIII.5.1. *Let f be continuous on R, and bounded. Define the convolution $P_y * f$ by*

$$P_y * f(x) = \int_{-\infty}^{\infty} P_y(x-t)f(t)dt.$$

Prove that $P_y * f(x)$ converges to $f(x)$ as $y \to 0$ for each x where f is continuous.

The proof should apply to the case when f is bounded, and continuous except at a finite number of points, etc.

Solution. Let x be a point where f is continuous. Changing variables and using the properties of the family $\{P_y\}$ we find

$$P_y * f(x) - f(x) = \int_{-\infty}^{\infty} P_y(t)f(x-t)dt - \int_{-\infty}^{\infty} P_y(t)f(x)dt$$

thus

$$|P_y * f(x) - f(x)| \le \int_{-\infty}^{\infty} P_y(t)|f(x-t) - f(x)|dt.$$

Let $\epsilon > 0$. Since f is continuous at x there exists $\delta > 0$ such that if $|t| < \delta$ then $|f(x-t) - f(x)| < \epsilon$. Write

$$\int_{-\infty}^{\infty} P_y(t)|f(x-t) - f(x)|dt = \int_{-\infty}^{-\delta} + \int_{-\delta}^{\delta} + \int_{\delta}^{\infty} P_y(t)|f(x-t) - f(x)|dt.$$

Choose y_0 as in **DIR 3** so that the sum of the first and third integral is bounded by $2B\epsilon$ where B is a bound for f. The middle integral is estimated by ϵ so we see that $P_y * f(x) \to f(x)$ as $y \to 0$.

Exercise VIII.5.2. *Let*

$$P_y(t) = \frac{1}{\pi} \frac{y}{t^2 + y^2} \quad \text{for } y > 0.$$

Prove that $\{P_y\}$ is a Dirac family. It has no special name, like the Poisson family as discussed in the text, but it is classical.

Solution. The first condition for a Dirac family holds because $y > 0$. Let

$$I(a, b) = \int_a^b P_y(t)dt.$$

Then changing variables $t = yu$ we find

$$I(a, b) = \frac{1}{\pi} \int_{a/y}^{b/y} \frac{du}{1+u^2} = \frac{1}{\pi}[\arctan u]_{a/y}^{b/y}.$$

Letting $a \to -\infty$ and $b \to \infty$ we find

$$\int_{-\infty}^{\infty} P_y(t)dt = \frac{1}{\pi}\left(\frac{\pi}{2} - \frac{-\pi}{2}\right) = 1$$

so the second condition for a Dirac family is verified.

To see why the third property is verified we let $b \to \infty$ so that

$$\int_a^\infty P_y(t)dt = \frac{1}{\pi}\left(\frac{\pi}{2} - \arctan\left(\frac{a}{y}\right)\right).$$

Given $\epsilon, \delta > 0$ choose y_0 such that for all $0 < y < y_0$ we have

$$\arctan(\delta/y) > \frac{\pi}{2} - \frac{\epsilon\pi}{2}.$$

Then for these values of y we obtain the inequality

$$\int_\delta^\infty P_y(t)dt < \frac{\epsilon}{2},$$

so by symmetry we get

$$\int_{-\infty}^{-\delta} + \int_\delta^\infty P_y(t)dt < \epsilon.$$

Exercise VIII.5.3. *Define for all x and $y > 0$:*

$$F(x, y) = P_y * f(x).$$

Prove that F is harmonic. In fact show that the Laplace operator

$$\left(\frac{\partial}{\partial x}\right)^2 + \left(\frac{\partial}{\partial y}\right)^2$$

applied to

$$\frac{y}{(t - x)^2 + y^2}$$

yields 0.

You will have to differentiate under an integral sign, with the integral being taken over the real line. You can handle this in two ways.
(i) Work formally and assume that everything is OK.
(ii) Justify all the steps. In this case, you have to use a lemma like that proved in Chapter XV, §1.

Solution. The map

$$g_t(z) = \frac{1}{t - z}$$

is holomorphic on the upper half plane and a direct computation shows that

$$\text{Im}(g_t(z)) = \pi P_y(x - t),$$

hence $P_y(x - t)$ is harmonic.

Exercise VIII.5.4. *Let u be a bounded continuous function on the closure of the upper half plane (i.e., on the upper half plane and the real line). Assume also that* *u is harmonic on the upper half plane, and that there are constants $c > 0$ and $K > 0$ such that*

$$|u(t)| \leq K \frac{1}{|t|^c} \quad \text{for all } |t| \text{ sufficiently large.}$$

Using the Dirac family of the preceding exercise, prove that there exists an analytic function f on the upper half plane whose real part is u. [Hint: Recall the integral formula of Exercise 23 of Chapter VI, §3.]

Solution. Let

$$g(z) = \int_{-\infty}^{\infty} \frac{u(t)}{t - z} dt$$

for $z \in H$. We want to show that g is analytic. Let V be a compact subset of the upper half plane H. Then for $z \in V$, $|z|$ is bounded. There exists $B > 0$ such that if $|t| > B$, then $|t - z| \geq |t|/2$ for all $z \in V$ and $|u(t)| \leq K/|t|^c$. So if $B_1 > B$ we have

$$\int_{-\infty}^{-B_1} + \int_{B_1}^{\infty} \frac{|u(t)|}{|t - z|} dt \leq 2K \int_{-\infty}^{-B_1} + \int_{B_1}^{\infty} \frac{1}{|t|^{1+c}} dt,$$

which implies the uniform convergence of the integral on V. The differentiation lemma implies that g is analytic. So the function f defined on the upper half plane by $f(z) = g(z)/(\pi i)$ is also analytic and we see that

$$v = \text{Re}(f) = P_y * u.$$

From the convergence result of Exercise 2, we see that $v(x, 0) = u(x)$. Now we must show that $v = u$ in the upper half plane. This is achieved by going to the disc and using Theorem 1.4. First note that since $v = P_y * u$, we have putting absolute values in the integral defining the convolution and using the fact that $\int P_y(x) dx = 1$

$$\|v\|_\infty \leq \|u\|_\infty$$

where $\| \cdot \|_\infty$ denotes the sup norm. Hence v is also bounded on the upper half plane, and it suffices to show that if w is a bounded harmonic function on the upper half plane which extends continuously to $\mathbf{R}$ with value 0, then w is identically 0. Consider the isomorphism $h : D \to H$ given in Chapter VII. Then $w \circ h$ is harmonic on D. It is also bounded and continuous at the boundary except possibly at one point. Since $w = 0$ on the boundary where it is continuous, we conclude from Theorem 1.4 that $w \circ h = 0$ and therefore $w = 0$. This concludes the exercise.

Let u be a continuous function on an open set U. We say that u satisfies the **circle mean value property** *at a point $z_0 \in U$ if*

$$u(z_0) = \frac{1}{2\pi} \int_0^{2\pi} u(z_0 + re^{i\theta}) d\theta$$

*for all r > 0 sufficiently small (so that in particular the disc $\overline{D}(z_0, r)$ is contained in U). We say that u satisfies the **disc mean value property** at a point $z_0 \in U$ if*

$$u(z_0) = \frac{1}{\pi r^2} \int \int_{\overline{D}(z_0,r)} u \, dx \, dy,$$

for all r > 0 sufficiently small. We say that the function satisfies the mean value property (either one) on U if it satisfies this mean value property at every point of U. By Theorem 3.3 and Theorem 5.5 we know that u is harmonic if and only if u satisfies the circle mean value property.

Exercise VIII.5.5. *Prove that u is harmonic if and only if u satisfies the disc mean value property on U.*

Solution. If u is harmonic, then is satisfies the circle mean value property

$$u(z_0) = \frac{1}{2\pi} \int_0^{2\pi} u(z_0 + \rho e^{i\theta}) d\theta.$$

Multiplying both sides by ρ and integrating from 0 to r, we see that u satisfies the disc mean value property.

Conversely, suppose that u satisfies the disc mean value property. Using polar coordinates, the disc mean value property gives

$$u(z_0) = \frac{1}{\pi r^2} \int_0^r \int_0^{2\pi} \rho u(z_0 + \rho e^{i\theta}) d\theta d\rho,$$

so writing r^2 as an integral we get

$$\int_0^r \rho u(z_0) d\rho = \frac{1}{2\pi} \int_0^r \int_0^{2\pi} \rho u(z_0 + \rho e^{i\theta}) d\theta d\rho.$$

Differentiating with respect to r we find

$$r u(z_0) = \frac{1}{2\pi} \int_0^{2\pi} r u(z_0 + re^{i\theta}) d\theta,$$

so cancelling r we find that u satisfies the circle mean value property and is therefore harmonic.

Exercise VIII.5.6. *Let H^+ be the upper half plane. For $z \in H^+$ define the function*

$$h_z(\zeta) = \frac{1}{2\pi i} \left(\frac{1}{\zeta - z} - \frac{1}{\zeta - \overline{z}} \right) \quad \text{for } \zeta \in H^+.$$

Then h_z is analytic in H^+ except for a simple pole at z. Let f be an analytic function on $H^+ \cup \mathbf{R}$ (i.e., on the closure of the upper half plane, meaning on an open set containing this closure). Suppose that f is bounded on $H^+ \cup \mathbf{R}$ Prove that

$$\int_{-\infty}^{\infty} f(t) h_z(t) \, dt = f(z).$$

This is the analogue of Theorem 4.1 for the upper half plane. [Hint: Integrate over the standard region from the calculus of residues, namely over the interval $[-R, R]$ and over the semicircle of radius R.]

Solution. Let $z \in H^+$. The function $f(\zeta)h_z(\zeta)$ is analytic on the closure of the upper half plane and has a simple pole at z with residue $f(z)/(2\pi i)$. For R large, we will have

$$\int_{\gamma_R} f(\zeta)h_z(\zeta)d\zeta = 2\pi i \frac{f(z)}{2\pi i},$$

where γ_R denotes the standard path consisting of the interval $[-R, R]$ and the semicircle $|z| = R$ in the upper half plane. To conclude the proof, we must show that the integral over the semicircle tends to 0. But this follows because f is bounded and

$$|h_z(\zeta)| \leq C \frac{|z - \bar{z}|}{|\zeta - z||\zeta - \bar{z}|} \leq \frac{B}{|\zeta|^2}.$$

Exercise VIII.5.7. *In Exercise 6, consider the case $z = i$. Let*

$$w = \frac{z - i}{z + i} \quad so \quad z = -i\frac{w + 1}{w - 1}$$

be the standard isomorphism between the upper half plane and the unit disc D. Show that

$$\left(\frac{1}{z - i} - \frac{1}{z + i}\right) dz = \frac{dw}{w}.$$

In light of Exercise 1, this shows that the kernel function in Exercise 2 corresponds to the Poisson kernel under the isomorphism between H^+ and D.

Solution. This is just a computation

$$\frac{1}{w}\frac{dw}{dz} = \frac{1}{w}\left(\frac{z + i - (z - i)}{(z + i)^2}\right)$$

$$= \frac{z + i}{z - i}\left(\frac{2i}{(z + i)^2}\right)$$

$$= \frac{2i}{(z + i)(z - i)}$$

$$= \frac{1}{z - i} - \frac{1}{z + i}.$$

Exercise VIII.5.8. *Let $x = r \cos\theta$, $y = r \sin\theta$ be the formulas for the polar coordinates. Let*

$$f(x, y) = f(r \cos\theta, r \sin\theta) = g(r, \theta).$$

Show that

$$\frac{\partial^2 g}{\partial r^2} + \frac{1}{r}\frac{\partial g}{\partial r} + \frac{1}{r^2}\frac{\partial^2 g}{\partial\theta^2} = \frac{\partial^2 f}{\partial x^2} + \frac{\partial^2 f}{\partial y^2}.$$

For the proof, start with the formulas

$$\frac{\partial g}{\partial r} = (D_1 f)\cos\theta + (D_2 f)\sin\theta \quad and \quad \frac{\partial g}{\partial \theta} = -(D_1 f)r\sin\theta + (D_2 f)r\cos\theta,$$

and take further derivatives with respect to r and with respect to θ, using the rule for derivative of a product, together with the chain rule. Then add the expression you obtain to form the left hand side of the relation you are supposed to prove. There should be enough cancellation on the right hand side to prove the desired relation.

Solution. The hint gives the proof away. For the details of this computation you can look at Exercise XII.3.5 in my book *Problems and Solutions for Undergraduate Analysis* published by Springer-Verlag.

Exercise VIII.5.9. *(a) For $t > 0$, let*

$$K(t, x) = K_t(x) = \frac{1}{\sqrt{4\pi t}}e^{-x^2/4t}.$$

*Prove that $\{K_t\}$ for $t \to 0$ is a Dirac family indexed by t, and $t \to 0$ instead of $n \to \infty$. One calls K the **heat kernel**.*
*(b) Let $D = (\partial/\partial x)^2 - \partial/\partial t$. Then D is called the **heat operator** (just as we defined a Laplace operator Δ). Show that $DK = 0$. (This is the analogue of the statement that $\Delta P = 0$ if P is the Poisson kernel.)*
*(c) Let f be a piecewise continuous bounded function on **R**. Let $F(t, x) = (K_t * f)(x)$. Show that $DF = 0$, i.e., F satisfies the heat equation.*

Solution. For $t > 0$ we have $K(t, x) \geq 0$. Changing variables $2\sqrt{t}u = x$ we find that

$$\int_{-\infty}^{\infty} K(t, x)dx = \int_{-\infty}^{\infty} \frac{1}{\sqrt{4\pi t}}e^{-x^2/4t}dx = \frac{1}{\sqrt{4\pi t}}\int_{-\infty}^{\infty} 2\sqrt{t}e^{-u^2}du = 1$$

because $\int_{-\infty}^{\infty} e^{-u^2}du = \sqrt{\pi}$.

We now show that the last property of a Dirac family holds. Let $\epsilon, \delta > 0$ be given. We see after the change of variable $2\sqrt{t}u = x$ that

$$\int_{\delta}^{\infty} K(t, x)dx = \frac{1}{\sqrt{\pi}}\int_{\delta/2\sqrt{t}}^{\infty} e^{-u^2}du.$$

But for all t close to 0, we have $\delta/2\sqrt{t} > 0$, and for $u > 1$ we have $e^{-u^2} \leq e^{-u}$, hence for all t close to 0 we have

$$\int_{\delta}^{\infty} K(t, x)dx = \frac{1}{\sqrt{\pi}}\int_{\delta/2\sqrt{t}}^{\infty} e^{-u^2}du \leq \frac{1}{\sqrt{\pi}}\int_{\delta/2\sqrt{t}}^{\infty} e^{-u}du \leq \frac{e^{-\delta/2\sqrt{t}}}{\sqrt{\pi}}.$$

Since $e^{-\delta/2\sqrt{t}} \to 0$ as $t \to 0$ and $K(t, x)$ is even in x, there exists $t_0 > 0$ such that for all $0 < t < t_0$ we have

$$\int_{-\infty}^{-\delta} + \int_{\delta}^{\infty} K(t, x)dx < \epsilon.$$

This concludes the proof that $\{K_t\}$ is a Dirac family for $t \to 0$.
(b) Differentiating once with respect to x we get

$$\frac{\partial}{\partial x} K(t, x) = \frac{1}{\sqrt{4\pi t}} \frac{-2x}{4t} e^{-x^2/4t}$$

so

$$\frac{\partial^2}{\partial x^2} K(t, x) = \frac{-1}{\sqrt{4\pi}} e^{-x^2/4t} \left[\frac{1}{2t^{3/2}} - \frac{x^2}{4t^{3/2}} \right].$$

Differentiating once with respect to t we find the same expression as $\partial^2 K / \partial x^2$ so $DK = 0$.
(c) By definition we have

$$F(t, x) = \int_{-\infty}^{\infty} K_t(x - y) f(y) dy.$$

Consider the integrand as a function of t and y. Suppose that t is in a compact interval. Then from the result found in (b), the fact that e^{-y^2} is rapidly decreasing and integrable and that f is bounded, we see that there exists integrable functions φ and ψ such that

$$|K_t(x - y) f(y)| \leq \varphi(y) \quad \text{and} \quad \left| \frac{\partial}{\partial t} K_t(x - y) f(y) \right| \leq \psi(y)$$

for all t and y. So we can differentiate under the integral. The same argument applied to $K_t(x - y) f(y)$ and $\partial / \partial x K_t(x - y) f(y)$, viewed as functions of x and y, yield

$$DF(t, x) = \int_{-\infty}^{\infty} DK_t(x - y) f(y) dy = (DK_t) * f(x) = 0.$$

IX

Schwarz Reflection

IX.2 Reflection Across Analytic Arcs

Exercise IX.2.1. *Let C be an arc of the unit circle $|z| = 1$, and let U be an open set inside the circle, having that arc as piece of its boundary. If f is analytic on U if f maps U into the upper half plane, f is continuous on C, and takes real values on C, show that f can be continued across C by the relation*

$$f(z) = \overline{f(1/\overline{z})}.$$

Solution. By Theorem 2.2, we know that f can be continued analytically across the arc. Thus the only thing to prove is that this continuation is given by the stated formula:

$$f(z) = \overline{f(1/\overline{z})}, \tag{1}$$

If z is on the unit circle, then $z = 1/\overline{z}$, and since f is real valued on the arc of the unit circle, the expression on the right of (1) is indeed equal to $f(z)$. Furthermore the expression on the right of (1) is continuous on the arc, because it is the composite of continuous functions. Therefore it suffices to show that the right side of the above equation is analytic for $|z| > 1$. Here we give one way of doing this, for another method, see the next exercise. We apply $\partial/\partial\overline{z}$ to the expression on the right and we get

$$\frac{\partial}{\partial\overline{z}} \overline{f(1/\overline{z})} = \overline{\frac{\partial}{\partial z} f(1/\overline{z})}.$$

By the chain rule we get

$$\frac{\partial}{\partial z} f(1/\bar{z}) = f'(1/\bar{z})\frac{\partial}{\partial z}(1/\bar{z}) = 0$$

because for instance

$$\frac{\partial}{\partial z}\frac{1}{\bar{z}} = \overline{\frac{\partial}{\partial \bar{z}}\frac{1}{z}} = 0$$

since $1/z$ is analytic. Then the expression on the right of (1) defines an analytic function for $|z| > 1$ such that $1/\bar{z} \in U$. Since it coincides with f on the real segment, it must be the same as the analytic function coming from Theorem 2.2, and this concludes the proof.

Exercise IX.2.2. *Suppose, on the other hand, that instead of taking real values on C, f takes on values on the unit circle, that is,*

$$|f(z)| = 1 \quad \text{for } z \text{ on } C.$$

Show that the analytic continuation of f across C is now given by

$$f(z) = 1/\overline{f(1/\bar{z})}.$$

Solution. We argue like in Exercise 1. Suppose that $|z| = 1$. Then $1/\bar{z} = z$ and by assumption $|f(z)|^2 = 1$, so $f(z) = 1/\overline{f(z)} = 1/\overline{f(1/\bar{z})}$ on the arc. By continuity of f there exists an open subset V of U such that $f \neq 0$ on V and $C \subset \partial V$. It is sufficient to show that $1/\overline{f(1/\bar{z})}$ is analytic for $|z| > 1$ and $1/\bar{z} \in V$. Let $|z_0| > 1$ and $1/\bar{z_0} \in V$. Put $w = 1/\bar{z}$ for z near z_0. Then f has a power series expansion

$$f(w) = \sum a_n(w - w_0)^n = \sum a_n \left(\frac{1}{\bar{z}} - \frac{1}{\bar{z_0}}\right).$$

Hence

$$\overline{f(1/\bar{z})} = \sum \bar{a}_n\left(\frac{1}{z} - \frac{1}{z_0}\right) = \sum \bar{a}_n\frac{(z_0 - z)^n}{(zz_0)^n}.$$

The series on the right is absolutely and uniformly convergent for z near z_0 whence it defines an analytic function of z. Since f is nonzero near w_0 we conclude that $1/\overline{f(1/\bar{z})}$ is analytic near z_0. This ends the proof.

Exercise IX.2.3. *Let f be a function which is continuous on the closed unit disc and analytic on the open disc. Assume that $|f(z)| = 1$ whenever $|z| = 1$. Show that the function f can be extended to a meromorphic function, with at most a finite number of poles in the whole plane.*

Solution. The function f can only have finitely many zeros, otherwise they would accumulate in $\bar{D}$ and this would contradict the fact that $|f(z)| = 1$ on the unit circle and that f extends continuously to the closed unit disc. Let $z_1, \ldots, z_m$ be the zeros of f inside the disc (not counting multiplicity) and let

$$m_i = \text{ord}_{z_i} f \quad \text{and} \quad P_i(z) = \frac{z_i - z}{1 - \bar{z_i}z}.$$

Then P_i has a zero of order 1 at z_i, so if we define

$$g(z) = \prod P_i(z)^{m_i}$$

then $h(z) = f(z)/g(z)$ is analytic in the unit disc, extends continuously to the unit circle and $|h(z)| = 1$ whenever $|z| = 1$ because $|P_i(z)| = 1$ whenever $|z| = 1$. By Theorem 2.2 we can extend h to an entire function $\tilde{h}$. If we then consider

$$\tilde{f}(z) = \tilde{h}(z)g(z)$$

we see that $\tilde{f} = f$ on the unit disc and $\tilde{f}$ is meromorphic with at most finitely many pole, namely at the points $1/\overline{z_i}$.

Exercise IX.2.4. *Let f be a meromorphic function on the open unit disc and assume that f has a continuous extension to the boundary circle. Assume also that f has only a finite number of poles in the unit disc, and that $|f(z)| = 1$ whenever $|z| = 1$. Prove that f is a rational function.*

Solution. Let $\{z_i\}$, $i = 1, \ldots, m$ be the zeros or poles of f inside the disc. Let

$$m_i = \operatorname{ord}_{z_i} f \quad \text{and} \quad P_i(z) = \frac{z_i - z}{1 - \overline{z}_i z}.$$

Then P_i has a zero of order 1 at z_i. Let

$$g(z) = \prod P_i(z)^{m_i} \quad \text{and} \quad h(z) = f(z)/g(z).$$

Then h has no zero or pole in D, h extends continuously to the closure of D, and since $|P_i(z)| = 1$ when $|z| = 1$, it follows that $|h(z)| = 1$ when $|z| = 1$. Hence by Exercise 2, h extends to an entire function by reflection in the unit circle. But there exists $\delta > 0$ such that $|h(z)| \geq \delta$ for $z \in \overline{D}$ because h has no zero in the closed disc. From the reflection, it follows that the analytic extension of h to $\mathbf{C}$ is bounded, whence h is constant by Liouville's theorem. This proves that $f = cg$ for some constant c, as was to be shown. Note that we could also use Exercise VI.1.34 to see that h is constant.

Exercise IX.2.5. *Work out the exercise left for you in the text, that is:*

Let W be an open neighborhood of a real interval $[a, b]$. Let g be analytic on W, and assume that $g'(t) \neq 0$ for all $t \in [a, b]$, and g is injective on $[a, b]$. Then there exists an open subset W_0 of W containing $[a, b]$ such that g is an analytic isomorphism of W_0 with its image.

[Hint: First, by compactness, show that there is some neighborhood of $[a, b]$ on which g' does not vanish, and so g is a local isomorphism at each point of this neighborhood. Let $\{W_n\}$ be a sequence of open sets shrinking to $[a, b]$, for instance the set of points at distance $< 1/n$ from $[a, b]$. Suppose g is not injective on each W_n. Let $z_n \neq z'_n$ be two points in W_n such that $g(z_n) = g(z'_n)$. The sequences $\{z_n\}$ and $\{z'_n\}$ have convergent subsequences, to points on $[a, b]$. If these limit points are distinct, this contradicts the injectivity of g on the real interval. If these limit points are equal, then for large n, the points are close to a point on the interval

and this contradicts the fact that g is a local isomorphism at each point of the interval.]

Solution. Around any point x on the interval, there exists, by continuity, an open neighborhood V_x on which $g' \neq 0$. Then $\bigcup V_x$ covers $[a, b]$ and by compactness we may choose a subcovering, which will be our desired open neighborhood. The rest of the proof is given in the hint.

Exercise IX.2.6. *Let U be an open connected set. Let f be analytic on U, and suppose f extends continuously to a proper analytic arc on the boundary of U, and this extension has value 0 on the arc. Show that $f = 0$ on U.*

Solution. Let $\gamma : [a, b] \to \mathbf{C}$ be the proper analytic arc. Since U is connected, it suffices to show that f is 0 on a neighborhood of some point which lies on the analytic arc. By Exercise 5, γ defines an analytic isomorphism of some neighborhood W of $[a, b]$ onto its image. So we may assume that $[a, b]$ is the boundary of some small open connected set V contained in the plane. Let V^+ and V^- denote the part of V lying above and below the real axis respectively. We wish to show that some small segment in $[a, b]$ is the boundary of either V^+ or V^-. Suppose some point z on (a, b) is not on the boundary of V^+. Then z is contained in some small open disc $D(z, r)$ which does not intersect V^+. Since every point on $[a, b]$ is a boundary point of V, we conclude that the full segment $[a, b] \cap D(z, r/2)$ is contained in the boundary of V^-. So assume without loss of generality that some segment I of $[a, b]$ is contained in the boundary of V^-. Since $f = 0$ on $[a, b]$ we can extend $f \circ \gamma$ analytically past I by Schwarz reflection and we then conclude that f is identically 0 on V^-. Since U is connected, we conclude that f is identically 0 on U.

X

The Riemann Mapping Theorem

X.1   Statement of the Theorem

Exercise X.1.1. *Let U be a simply connected open set. Let z_1, z_2 be two points of U. Prove that there exists a holomorphic automorphism f of U such that $f(z_1) = z_2$. (Distinguish the cases when $U = \mathbf{C}$ and $U \neq \mathbf{C}$.)*

Solution. Suppose $U = \mathbf{C}$. Then $f(z) = z + (z_2 - z_1)$ is a solution to the problem. If $U \neq \mathbf{C}$, then by the Riemann mapping theorem, we can map U to the open unit disc by a holomorphic isomorphism $g : U \to D$. The automorphisms of the disc act transitively, so we can find $f \in \operatorname{Aut}(D)$ which maps $g(z_1)$ to $g(z_2)$. Then $g^{-1} \circ f \circ g$ is a solution to the problem.

Exercise X.1.2. *Let $f(z) = 2z/(1 - z^2)$. Show that f gives an isomorphism of the shaded region with a half disc. Describe the effect of f on the boundary. See the figure on page 180.*
   What is the effect of f on the reflection of the region across the y-axis?

Solution. Let U be the shaded region and let γ be its boundary. Let D_L be the left half unit disc. We first show that $f(\gamma) = \partial D_L$. We use polar coordinates to parametrize the arc included in γ. Let $u(\theta) = 1 + \sqrt{2}e^{i\theta}$ with $3\pi/4 \leq \theta \leq 5\pi/4$. We have

$$f(u(\theta)) = \frac{2(1 + \sqrt{2}e^{i\theta})}{(1 + 1 + \sqrt{2}e^{i\theta})(1 - 1 - \sqrt{2}e^{i\theta})} = -\frac{\sqrt{2}e^{-i\theta} + 2}{2 + \sqrt{2}e^{i\theta}}.$$

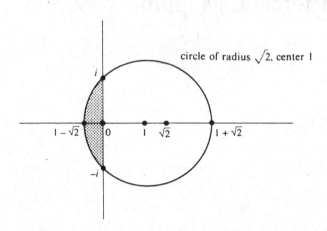

circle of radius $\sqrt{2}$, center 1

But $\overline{2 + \sqrt{2}e^{i\theta}} = \sqrt{2}e^{-i\theta} + 2$ so $|f(u(\theta))| = 1$ which proves that the image of γ under f is contained in the unit circle. But

$$\arg f(u(\theta)) = \arg(\sqrt{2}e^{-i\theta} + 2) - \arg(2 + \sqrt{2}e^{i\theta}) + \pi$$
$$= \pi - 2\arg(2 + \sqrt{2}e^{i\theta}),$$

and $2 + \sqrt{2}e^{i\theta}$ parametrizes the circle centered at 2 or radius $\sqrt{2}$, so from the figure

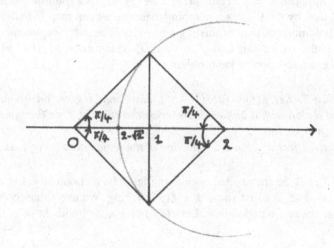

we see that the range of $\arg f(u(\theta))$ is $[\pi/2, 3\pi/2]$. Now we describe the effect of f on the vertical segment iy with $-1 \le y \le 1$. We have

$$f(iy) = \frac{2iy}{1-(iy)^2} = i\frac{2y}{1+y^2},$$

and since $y \mapsto (2y)/(1+y^2)$ is a bijection from $[-1, 1]$ onto itself, the vertical segment is mapped onto $i[-1, 1]$. This proves that $f(\gamma) = \partial D_L$.

Theorem 4.3 of Chapter VII shows that f is an isomorphism of U with D_L. Now to see what happens to the reflection of the region U across the y-axis, note that the new region is also the image of the region U under the symmetry of center the origin. Then it suffices to see that

$$f(-z) = -f(z)$$

to conclude that the image of the new region is the reflection of $f(U)$ across the y-axis, whence the right half disc.

X.2   Compact Sets in Function Spaces

Let U be an open set, and let $\{K_s\}$ ($s = 1, 2, \ldots$) be a sequence of compact subsets of U such that K_s is contained in the interior of K_{s+1} for all s, and the union of the sets K_s is U. For f holomorphic on U, define

$$\sigma_s(f) = \min(1, \|f\|_s),$$

where $\|f\|_s$ is the sup norm of f on K_s. Define

$$\sigma(f) = \sum_{s=1}^{\infty} \frac{1}{2^s}\sigma_s(f).$$

Exercise X.2.1. *Prove that σ satisfies the triangle inequality on $\mathrm{Hol}(U)$, and defines a metric on $\mathrm{Hol}(U)$.*

Solution. We show that σ defines a metric d on $\mathrm{Hol}(U)$ given by

$$d(f, g) = \sigma(f - g).$$

The only non trivial property to verify is the triangle inequality. First we show that the triangle inequality holds for σ. It suffices to show that the triangle inequality holds for σ_s for all s.

Suppose that $\|f + g\|_s \le 1$. If one of $\|f\|_s$ or $\|g\|_s$ is ≥ 1, then

$$\sigma_s(f + g) = \|f + g\|_s \le 1 \le \sigma_s(f) + \sigma_s(g),$$

and if not, then

$$\sigma_s(f + g) = \|f + g\|_s \le \|f\|_s + \|g\|_s \le \sigma_s(f) + \sigma_s(g).$$

Suppose that $\|f + g\|_s > 1$, then

$$\sigma_s(f + g) = 1 < \|f + g\|_s \le \|f\|_s + \|g\|_s$$

but $\sigma_s(f) + \sigma_s(g)$ is either ≥ 1 or equal to $\|f\|_s + \|g\|_s$, so this proves that the triangle inequality holds for σ_s. Thus σ satisfies the triangle inequality and

$$d(f, h) = \sigma(f - h) \leq \sigma(f - g) + \sigma(g - h) = d(f, g) + d(g, h),$$

thereby concluding the proof that d is a metric.

Exercise X.2.2. *Prove that a sequence $\{f_n\}$ in $\mathrm{Hol}(U)$ converges uniformly on every compact subset of U if and only if it converges for the metric σ.*

Solution. Suppose that $f_n \to f$ uniformly on every compact subset of U. Given $0 < \epsilon < 1$, choose an integer n_0 such that $1/2^{n_0+1} < \epsilon$. Select an integer N such that $N > n_0$ and $\sigma_{n_0}(f_n - f) < \epsilon$ for all $n \geq N$. Since $K_s \subset K_{s+1}$ we have $\sigma_s \leq \sigma_{s+1}$ and therefore for all $n \geq N$ we find that

$$\sigma(f_n - f) = \sum_{s=1}^{n_0} \frac{1}{2^s}\sigma_s(f_n - f) + \sum_{n_0+1}^{\infty} \frac{1}{2^s}\sigma_s(f_n - f)$$

$$\leq \sum_{s=1}^{n_0} \frac{1}{2^s}\sigma_s(f_n - f) + \sum_{n_0+1}^{\infty} \frac{1}{2^s}$$

$$\leq \sigma_{n_0}(f_n - f) + \frac{1}{2^{n_0+1}}$$

$$\leq 2\epsilon.$$

Conversely, assume that $f_n \to f$ for the metric d. Let K be a compact set and let $0 < \epsilon < 1$. Select and integer n_0 such that for all $n \geq n_0$ we have $K \subset K_n$. Select N such that for all $n \geq N$ we have $\sigma(f_n - f) < \epsilon/2^{n_0}$. So for all $n \geq N$ we have

$$\frac{1}{2^{n_0}}\sigma_{n_0}(f_n - f) \leq \sum_{s=1}^{\infty} \frac{1}{2^s}\sigma_s(f_n - f) < \frac{\epsilon}{2^{n_0}},$$

hence $\sigma_{n_0}(f_n - f) < \epsilon$ for all $n \geq N$. This implies that $f_n \to f$ uniformly on K because

$$\|f_n - f\|_K \leq \|f_n - f\|_{n_0} = \sigma_{n_0}(f_n - f) < \epsilon$$

for all $n \geq N$.

Exercise X.2.3. *Prove that $\mathrm{Hol}(U)$ is complete under the metric σ.*

Solution. Let $\{f_n\}_{n=1}^{\infty} \subset \mathrm{Hol}(U)$ be a Cauchy sequence for the metric d induced by σ (see Exercise 1 of this section). For each s we have

$$\frac{1}{2^s}\sigma_s(f_n - f_m) \leq \sigma(f_n - f_m)$$

so $\{f_n\}_{n=1}^{\infty}$ is Cauchy for the sup norm on every compact set. Hence $\{f_n\}_{n=1}^{\infty}$ converges to a limit function f and the convergence is uniform on every compact set therefore f is holomorphic on U. By Exercise 2 we know that $f_n \to f$ for d so $\mathrm{Hol}(U)$ is complete for the metric d. This completes the proof.

Exercise X.2.4. *Prove that the map* $f \mapsto f'$ *is a continuous map of* Hol(U) *into itself, for the metric* σ.

Solution. Let d be the metric induced by σ (see Exercise 1). In a metric space, continuity can be verified by looking at sequences. We use Exercise 2 repeatedly. Suppose that $f_n \to f$ with respect to d. Then $f_n \to f$ uniformly on every compact subset of U, and it is a corollary of Cauchy's formula that $f_n' \to f'$ uniformly on every compact subset of U. Hence $f_n' \to f'$ with respect to d. This proves that the map $f \mapsto f'$ is continuous for the metric d.

Exercise X.2.5. *Show that a subset of* Hol(U) *is relatively compact in the sense defined in the text if and only if it is relatively compact with respect to the metric* σ *in the usual sense, namely its closure is compact.*

Solution. In this exercise we use repeatedly the fact that in a metric space, compactness is equivalent to sequential compactness.

Let Φ be a subset of Hol(U). By definition Φ is relatively compact if every sequence in Φ has a subsequence which converges uniformly on every compact subset of U. Suppose Φ is relatively compact according to this definition. We want to show that $\overline{\Phi}$ is compact in the metric space (Hol$(U), d$), where d is the metric induced by σ. Let $\{f_n\}_{n=1}^{\infty}$ be a sequence in $\overline{\Phi}$ and for each n choose a function $g_n \in \Phi$ such that $d(f_n, g_n) < 1/n$. Then by assumption, $\{g_n\}$ has a subsequence $\{g_{n_k}\}$ which converges uniformly on compact subsets of U. By Exercise 2, $\{g_{n_k}\}$ converges with respect to d and therefore $\{f_{n_k}\}$ converges with respect to d. Since $\overline{\Phi}$ is closed we conclude that $\{f_n\}$ has a subsequence which converges in $\overline{\Phi}$, as was to be shown.

Conversely, suppose that Φ is relatively compact with respect to the metric d induced by σ. Let $\{f_n\}$ be a sequence in Φ. Then $\overline{\Phi}$ is sequentially compact, so $\{f_n\}$ has a subsequence which converges with respect to d. Exercise 2 implies that this subsequence converges uniformly on every compact subset of U, so Φ is relatively compact.

Exercise X.2.6. *Let* Φ *be the family of all analytic functions*

$$f(z) = z + a_2 z^2 + a_3 z^3 + \cdots$$

on the open unit disc, such that $|a_n| \leq n$ *for each* n. *Show that* Φ *is relatively compact.*

Solution. By assumption $\Phi \subset$ Hol(U). To apply Theorem 2.1 (Ascoli's theorem) we must show that Φ is uniformly bounded on compact sets in the unit disc. Let K be a compact set in D. There exits a positive number $c < 1$ such that for all $z \in K$ we have $|z| \leq c$. If $f \in \Phi$ we get the following estimate

$$|f(z)| \leq \sum |a_n| c^n \leq \sum n c^n.$$

Applying the ratio test we find that $\sum n c^n < \infty$. This implies that Φ is uniformly bounded on compact sets and we are done.

Exercise X.2.7. *Let $\{f_n\}$ be a sequence of analytic functions on U, uniformly bounded. Assume that for each $z \in U$ the sequence $\{f_n(z)\}$ converges. Show that $\{f_n\}$ converges uniformly on compact subsets of U.*

Solution. Since $\{f_n\}$ is uniformly bounded, the beginning of the proof of Theorem 2.1 shows that the family $\{f_n\}$ is equicontinuous on compact sets. Let $\epsilon > 0$ and K be a compact subset of U. Choose $\delta > 0$ such that for all $|z - z'| < \delta$ and $z, z' \in K$ we have $|f_n(z) - f_n(z')| < \epsilon$ for all n. From the covering $\bigcup_{z \in K} B(z, \delta)$ of K, choose a finite subcovering $B(z_1, \delta), \ldots, B(z_l, \delta)$. Choose N so large that

$$|f_n(z_i) - f_m(z_i)| < \epsilon$$

for all $n, m > N$ and all i. Now we show that $\{f_n\}$ is uniformly convergent on K. Given $z \in K$ choose i such that $z \in B(z_i, \delta)$. Then for all $n, m > N$ we have

$$|f_n(z) - f_m(z)| \le |f_n(z) - f_n(z_i)| + |f_n(z_i) - f_m(z_i)| + |f_m(z_i) - f_m(z)|$$

$$\le \epsilon + \epsilon + \epsilon.$$

This concludes the proof.

XI

Analytic Continuation along Curves

XI.1 Continuation Along a Curve

Exercise XI.1.1. *Let f be analytic in the neighborhood of a point z_0. Let k be a positive integer, and let $P(T_0, \ldots, T_k)$ be a polynomial in $k+1$ variables. Assume that*

$$P(f, Df, \ldots, D^k f) = 0,$$

where $D = d/dz$. If f can be continued along a path γ, show that

$$P(f_\gamma, Df_\gamma, \ldots, D^k f_\gamma) = 0.$$

Solution. Let $(f_0, D_0), \ldots, (f_n, D_n)$ be an analytic continuation along γ of the function f. So $f_n = f_\gamma$ in a neighborhood of the endpoint of γ. Since $f_0 = f_1$ on $D_0 \cap D_1$ we have $P(f_1, Df_1, \ldots, D^k f_1) = 0$ on $D_0 \cap D_1$. But $P(f_1, Df_1, \ldots, D^k f_1)$ is analytic on the connected open set D_1 so $P(f_1, Df_1, \ldots, D^k f_1) = 0$ on D_1. By induction we get $P(f_\gamma, Df_\gamma, \ldots, D^k f_\gamma) = 0$, as was to be shown.

Exercise XI.1.2 (Weierstrass). *Prove that the function*

$$f(z) = \sum z^{n!}$$

cannot be analytically continued to any open set strictly larger than the unit disc. [Hint: If z tends to 1 on the real axis, the series clearly becomes infinite. Rotate z by a k-th root of unity for positive integers k to see that the function becomes infinite on a dense set of points on the unit circle.]

Solution. The set $\{e^{2\pi i r}\}_{r \in \mathbf{Q}}$ is dense in the unit circle. Let $z_{p,q} = e^{2\pi i (p/q)}$ where p and q are integers with $q > 0$. Then for $0 < t < 1$ we have

$$f(tz_{p,q}) = \sum_{n=1}^{q-1}(tz_{p,q})^{n!} + \sum_{n=q}^{\infty}(tz_{p,q})^{n!}$$

$$= \sum_{n=1}^{q-1}(tz_{p,q})^{n!} + t^{q!} + t^{(q+1)!} + \cdots$$

because $(z_{p,q})^{n!} = 1$ for $n \geq q$. Thus $\lim_{t \to 1}|f(tz_{p,q})| = \infty$ for all p, q integers with $q > 0$, and this proves that f cannot be analytically continued to any open set strictly larger than the disc.

Exercise XI.1.3. *Let U be a connected open set and let u be a harmonic function on U. Let D be a disc contained in U and let f be an analytic function on D such that $u = \mathrm{Re}(f)$. Show that f can be analytically continued along every path in U.*

Solution. Let $\gamma : [a, b] \to U$ be a curve in U. By compactness of $\gamma([a, b])$ we can find a partition $a = a_0 < a_1 < \cdots < a_{n+1} = b$ and discs $D_0, \ldots, D_n$ such that $\gamma([a_i, a_{i+1}]) \subset D_i$, $\gamma(a_{i+1}) \in D_i \cap D_{i+1}$ and $D_i \subset U$. Also we can assume that $D_0 = D$. For each i, there exists an analytic function $\tilde{f}_i$ on D_i such that $\mathrm{Re}(\tilde{f}_i) = u$. On $D_0 \cap D_1$ the functions f and $\tilde{f}_1$ have the same real part so they differ by an imaginary constant, say $f - \tilde{f}_1 = iK_1$. If we define $f_1 = \tilde{f}_1 + iK_1$, then $f_1 = f$ on $D_0 \cap D_1$. By induction it is clear that f can be continued along γ.

Exercise XI.1.4. *Let U be a simply connected open set in $\mathbf{C}$ and let u be a real harmonic function on U. Reprove that there exists an analytic function f on U such that $u = \mathrm{Re}(f)$ by showing that if (f_0, D_0) is analytic on a disc and $\mathrm{Re}(f_0) = u$ on the disc, then (f_0, D_0) can be continued along every curve in U.*

Solution. Let $z_0 \in U$ and let D_0 be a disc containing z_0 and such that $D_0 \subset U$. We construct an analytic function on D_0 whose real part is u using the Poisson integral given in Chapter VIII

$$f_0(z) = \frac{1}{2\pi i} \int_{\partial D_0} \frac{\zeta + z}{\zeta - z} u(\zeta) \frac{d\zeta}{\zeta}.$$

We define $f = f_0$ in a neighborhood of z_0. Let $w \in U$ and let $\gamma : [a, b] \to U$ be a curve joining z_0 and w. Arguing like in Exercise 3, using the Poisson integral formula to define $\tilde{f}_i$ we see that f can be continued along γ, and in a neighborhood of w we get a function f_γ whose real part is u. Since U is simply connected, the Monodromy theorem applies, and we can define f unambiguously in a neighborhood of w by $f = f_\gamma$. This shows that we can define a global analytic function whose real part is u.

XI.2   The Dilogarithm

Exercise XI.2.1. *We investigate the analytic continuation of the dilogarithm for the curves illustrated in the figure. Let $z_1 = 1/2$.*

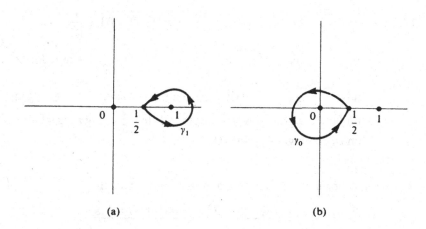

(a)   (b)

(a) Let γ_1 be a curve as shown on figure (a), circling 1 exactly once. How does the analytic continuation of L_2 along γ_1 differ from L_2 in a neighborhood of z_1?
(b) How does the analytic continuation of L_2 along the path γ_0 on figure (b) differ from L_2.?
(c) If you continue L_2 first around γ_0 and then around γ_1, how does this continuation differ from continuing L_2 first around γ_1 and then around γ_0? [They won't be equal!]

Solution. (a) By a calculation identical to the one on page 332 of Lang's book, we see that the analytic continuation of L_2 along γ_1 will differ from L_2 in a neighborhood of z_1 by $2\pi i \log_{pr}(1/2)$.
(b) Along the path γ_0, the difference will be 0 because $\log(1 - \zeta)/\zeta$ is holomorphic near the origin.
(c) Going around γ_0 first does not affect anything, so we just have a difference of $2\pi i \log_{pr}(1/2)$. Going around γ_1 first we pick up a pole at the origin with residue an integral multiple of $2\pi i$. So going around γ_0 we pick up this residue, so that we finally obtain a difference of $2\pi i \log_{pr}(1/2) + 2\pi i(2\pi i m)$ where m is some integer.

Exercise XI.2.2. *Let D be the Bloch–Wigner function.*
(a) Show that $D(1/z) = -D(z)$ for $z \neq 0$, and so D extends in a natural way to a continuous function on $\mathbf{C} \cup \{\infty\}$.
(b) Show that $D(z) = -D(1 - z)$.

Solution. (a) We claim that for x real positive there exists a real constant c such that

$$L(-1/x) = -L(-x) + c - \frac{1}{2}(\log x)^2.$$

If this is true, then we see that

$$D(1/z) = -\operatorname{Im} L(z) - \operatorname{Im}\left(\frac{1}{2}(\log(-z))^2\right) + \arg(1 - (1/z))\log\frac{1}{|z|}$$

$$= -\operatorname{Im} L(z) - \operatorname{Im}\left(\frac{1}{2}(\log(-z))^2\right) - \arg(z - 1))\log|z| + \arg z \log|z|$$

$$= -\operatorname{Im} L(z) - \frac{1}{2}(2\arg(-z)\log|z|) - \arg(z - 1)\log|z| + \arg z \log|z|$$

$$= -\operatorname{Im} L(z) - \arg(-z)\log|z| - \arg(z - 1)\log|z| + \arg z \log|z|$$

$$= -\operatorname{Im} L(z) - \arg(1 - z)\log|z|$$

$$= -D(z).$$

To prove the functional equation for the dilogarithm, we note that

$$\frac{d}{dx}L(-1/x) = \frac{\log(1 + 1/x)}{x} = \frac{\log(1 + x) - \log x}{x}.$$

Integrating we find the desired formula.

(b) We claim that

$$L(z) + L(1 - z) = c - \log z \log(1 - z)$$

where c is a real constant. This follows from integration by parts in the formula for $L(z)$. Then

$$D(z) + D(1 - z) = -\operatorname{Im}(\log z \log(1 - z)) + \arg(1 - z)\log|z| + \arg z \log|1 - z|$$

$$= -\arg(1 - z)\log|z| - \arg z \log|1 - z|$$

$$+ \arg(1 - z)\log|z| + \arg z \log|1 - z|.$$

Exercise XI.2.3. *For $k \geq 2$ define the **polylogarithm function***

$$L_k(z) = \sum_{n=1}^{\infty} \frac{z^n}{n^k} \quad \text{for } |z| < 1.$$

Show that for every positive integer N,

$$L_k(z^N) = N^{k-1} \sum_{\zeta^N=1} L_k(\zeta z).$$

where the sum is taken over all N-th roots of unity. [Hint: Observe that if ζ is an N-th root of unity, $\zeta \neq 1$, then

$$1 + \zeta + \cdots + \zeta^{N-1} = \frac{1 - \zeta^N}{1 - \zeta} = 0.]$$

Solution. Let $\omega = e^{2\pi i/N}$, so that the N-th roots of unity are $\omega^0, \omega^1, \dots, \omega^{N-1}$. Then

$$\sum_{\zeta^N=1} L_k(\zeta z) = \sum_{j=0}^{N-1} \sum_{n=1}^{\infty} \frac{(\omega^j z)^n}{n^k}$$

$$= \sum_{n=1}^{\infty} \frac{z^n}{n^k} \sum_{j=0}^{N-1} (\omega^n)^j.$$

Note that $\omega^n = 1$ precisely when $n = 0 \bmod N$, and when this is not the case we have

$$\sum_{j=0}^{N-1} (\omega^n)^j = \frac{1 - \omega^{nN}}{1 - \omega^n} = 0.$$

These observations imply that

$$\sum_{\zeta^N=1} L_k(\zeta z) = \sum_{l=1}^{\infty} \frac{z^{lN}}{(lN)^k} N$$

$$= \frac{1}{N^{k-1}} \sum_{l=1}^{\infty} \frac{(z^N)^l}{l^k}$$

$$= \frac{1}{N^{k-1}} L_k(z^N).$$

Exercise XI.2.4. *Prove the relation*

$$\prod_{\zeta^N=1} (1 - \zeta X) = 1 - X^N,$$

where the product is taken over all N-th roots of unity ζ.

Solution. Let $\omega = e^{2\pi i/N}$. Then, $\omega^0, \omega, \dots, \omega^{N-1}$ are N distinct N-th roots of unity, so

$$X^N - 1 = \prod_{j=0}^{N-1} (X - \omega^j). \tag{1}$$

Now observe that

$$\omega^{-0} \omega^{-1} \cdots \omega^{-(N-1)} = e^{-\frac{2\pi i}{N}(0+1+\cdots+(N-1))}$$

$$= e^{-\frac{2\pi i}{N} \frac{(N-1)N}{2}}$$

$$= e^{-\pi i(N-1)} = (-1)^{N-1},$$

so that multiplying the equation (1) by -1 we get

$$1 - X^N = -\prod_{j=0}^{N-1} (X - \omega^j)$$

$$= (-1)^N (-1)^{N-1} \prod_{j=0}^{N-1} (X - \omega^j)$$

$$= \prod_{j=0}^{N-1} -\omega^{-j}(X - \omega^j)$$

$$= \prod_{j=0}^{N-1} (1 - \omega^{-j} X)$$

$$= \prod_{\zeta^N = 1} (1 - \zeta X).$$

XII

Applications of the Maximum Modulus Principle and Jensen's Formula

XII.1   Jensen's Formula

Exercise XII.1.1. *Let f be analytic on the closed unit disc and assume that $|f(z)| \leq 1$ for all z in this set. Suppose also that $f(1/2) = f(i/2) = 0$. Prove that $|f(0)| \leq 1/4$.*

Solution. If $f(0) = 0$ there is nothing to prove. If not, we apply Theorem 1.1 which states that

$$|f(0)| \leq \frac{\|f\|_R}{R^N} |z_1 \cdots z_N|$$

where R is the radius of the disc and $z_1, \ldots, z_N$ the zeros of f ordered in increasing absolute value. Putting $R = 1$, and using the fact that $|f(z)| \leq 1$ on the closed unit disc, we obtain

$$|f(0)| \leq |z_1 \cdots z_N|.$$

We know the exact value of two zeros of f, and for the other zeros we have $|z_i| \leq 1$. Hence

$$|f(0)| \leq \frac{1}{2} \times \frac{1}{2} = \frac{1}{4}.$$

Exercise XII.1.2. *Let f be analytic on a disc $\overline{D}(z_0, R)$, and suppose f has at least n zeros in a disc $D(z_0, r)$ with $r < R$ (counting multiplicities). Assume $f(z_0) \neq 0$. Show that*

$$\left(\frac{R}{r}\right)^n \leq \|f\|_R / |f(z_0)|.$$

Solution. Define an analytic function g on $D(0, R)$ by $g(z) = f(z + z_0)$. Then $\|g\|_R = \|f\|_R$ where $\|\cdot\|_R$ denotes in the first case the sup norm on the disc $D(0, R)$ and in the second case the sup norm on $D(z_0, R)$. If $z_1, \ldots, z_n$ denotes n zeros of f in the disc $D(z_0, r)$ consider n zeros of g given by $z_1' = z_1 - z_0, \ldots, z_n' = z_1 - z_0$. Clearly, $z_1', \ldots, z_n' \in D(0, r)$ so by Theorem 1.1 we have

$$|g(0)| \le \frac{\|g\|_R}{R^n} |z_1' \cdots z_n'| \le \frac{\|g\|_R}{R^n} r^n.$$

Therefore

$$|f(z_0)| \le \|f\|_R \frac{r^n}{R^n}.$$

This concludes the proof.

Exercise XII.1.3. *Let f be an entire function. Write $z = x + iy$ as usual. Assume that for every pair of real numbers $x_0 < x_1$ there is a positive integer M such that $f(x + iy) = O(y^M)$ for $y \to \infty$, uniformly for $x_0 \le x \le x_1$. The implied constant in the estimate depends on x_0, x_1 and f. Let $a_1 < a_2$ be real numbers. Assume that $1/f$ is bounded on $\text{Re}(z) = a_2$. For $T > 0$, let $N_f(T)$ be the number of zeros of f in the box*

$$a_1 \le x \le a_2 \quad and \quad T \le y \le T + 1.$$

Prove that $N_f(T) = O(\log T)$ for $T \to \infty$. [Hint: Use an estimate as in Exercise 2 applied to a pair of circles centered at $a_2 + iy$ and of constant radius.]

Remark. *The estimate of Exercise 3 is used routinely in analytic number theory to estimate the number of zeros of a zeta function in a vertical strip.*

Solution. Let $T > 0$ and let $z_T = a_2 + iT$. We also denote by $B(T)$ the box described in the statement of the exercise. Select $r > 0$ such that $B(T)$ is contained in the disc of radius r centered at z_T, and let $R = 2r$. The picture is on the next page.

By Exercise 2 we have

$$\left(\frac{R}{r} \right)^{N_f(T)} \le \frac{\|f\|_R}{|f(z_T)|}.$$

Taking the log on both sides and using the fact that $R = 2r$ we get

$$N_f(T) \log 2 \le \log \|f\|_R - \log |f(z_T)|.$$

But $\overline{D}(z_T, R)$ is contained in some strip $x_0 \le x \le x_1$ and in this strip we have $f(x + iy) = O(y^M)$ for $y \to \infty$ uniformly in x, and R is independent of T (R is fixed) so we have $\|f\|_{C(z_T, R)} = O(T^m)$ as $T \to \infty$. By assumption, f is bounded from below on the line $\text{Re}(z) = a_2$ so there exists $\delta > 0$ such that $|f(z_T)| \ge \delta$ for all large T. Combining all these results with the above inequality, we obtain $N_f(T) = O(\log T)$ for $T \to \infty$, as was to be shown.

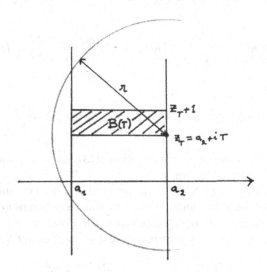

Next we develop extensions of the Poisson formula. We first set some notation. For $a \in D_R$, define

$$G_R(z, a) = G_{R,a}(z) = \frac{R^2 - \bar{a}z}{R(z - a)}.$$

Then $G_{R,a}$ has precisely one pole on $\overline{D}_R$ and no zeros. We have

$$|G_{R,a}(z)| = 1 \quad \text{for } |z| = R.$$

Exercise XII.1.4. *Apply the Poisson formula of Chapter VIII, §4 to prove the following theorem.*

Poisson–Jensen Formula. *Let f be meromorphic on $\overline{D}_R$. Let U be a simply connected open subset of D_R not containing the zeros or poles of f. Then there is a real constant K such that for z in this open set, we have*

$$\log f(z) = \int_0^{2\pi} \log |f(Re^{i\theta})| \frac{Re^{i\theta} + z}{Re^{i\theta} - z} \frac{d\theta}{2\pi} - \sum_{a \in D_R} (\text{ord}_a\, f) \log G_R(z, a) + iK.$$

[For the proof, assume first that f has no zeros and poles on the circle C_R. Let

$$h(z) = f(z) \prod G_R(z, a)^{\text{ord}_a\, f}$$

and apply Poisson to $\log h$. Then take care of the zeros and poles on C_R in the same way as in the Jensen formula.]

Solution. Suppose first that f has no zeros or poles on C_R. Define h by

$$h(z) = f(z) \prod_{a \in D_R} G_R(z, a)^{\text{ord}_a\, f}.$$

Then h has no zeros or poles on $\overline{D}_R$ and $\log h(z)$ is holomorphic on $\overline{D}_R$. Then by Poisson (Theorem 4.2 of Chapter VIII) there exists an imaginary constant iK such that

$$\log h(z) = \int_0^{2\pi} \operatorname{Re}\left(\log h(Re^{i\theta})\right) \frac{Re^{i\theta} + z}{Re^{i\theta} - z} \frac{d\theta}{2\pi} + iK,$$

so

$$\log f(z) + \sum_{a \in D_R} (\operatorname{ord}_a f) \log G_R(z, a) = \int_0^{2\pi} \log |f(Re^{i\theta})| \frac{Re^{i\theta} + z}{Re^{i\theta} - z} \frac{d\theta}{2\pi} + iK.$$

For the general case, we see that it is sufficient to prove the formula for a number z_0 on the circle, and the function $g(z) = z - z_0$. Then g is holomorphic on the closed disc $\overline{D}_R$ except at z_0. We give two ways to deal with the improper integral.

The first method uses the dominated convergence theorem of Lebesgue integration. Pick a sequence of number R' approaching R. For z in a closed disc $D_{R'}$ and R' sufficiently close to R, we have by Theorem 4.2 of Chapter VIII,

$$g(z) = \int_0^{2\pi} \log |R'e^{i\theta} - z_0| \frac{Re^{i\theta} + z}{Re^{i\theta} - z} \frac{d\theta}{2\pi} + iK,$$

where $K = \operatorname{Im}(\log(-z_0))$ for a definite choice of the branch of the logarithm. It remains to show that the integral on the right approaches the integral with R' replaced by R. This is an immediate consequence of the dominated convergence theorem, provided we can prove that for all R' sufficiently close to R and θ close to $\theta_0 = \arg z_0$, we have

$$\left|\log |R'e^{i\theta} - z_0|\right| \leq \left|\log |Re^{i\theta} - z_0|\right|.$$

After a rotation bringing z_0 to R and after setting $x = R/R'$, so x is real > 1, we are reduced to the case when $R = 1, z_0 = 1$. Thus we have to prove that for θ near 0 and $x > 1$ we have

$$|e^{i\theta} - 1| \leq |e^{i\theta} - x|,$$

and therefore

$$\left|\log |e^{i\theta} - x|\right| \leq \left|\log |e^{i\theta} - 1|\right|$$

which is clear from drawing a picture.

The second method uses the technique of kinks on the circle as for the Lemma in the proof of Jensen's formula. We do this as follows. Let $\gamma(\epsilon)$ be the circle of radius R modified to have a little kink around z_0 of radius ϵ. So $\gamma(\epsilon)$ consists of two curves. The first curve consists of al numbers $Re^{i\theta}$ with $|\theta - \theta_0| \geq \delta$, where δ depends on ϵ and tends to 0 as $\epsilon \to 0$. We let this curve be $\gamma_1(\epsilon)$. The second curve is an arc of a circle of radius ϵ around z_0, which we denote by $\gamma_2(\epsilon)$.

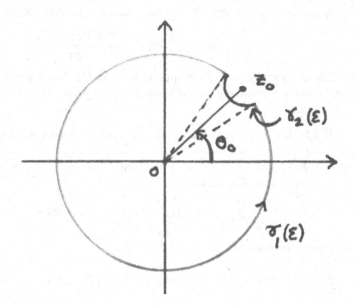

By Cauchy's theorem

$$\log(z - z_0) = g(z) = \frac{1}{2\pi i} \int_{\gamma(\epsilon)} \frac{g(\zeta)}{\zeta - z} d\zeta = \int_{\gamma_1(\epsilon)} \frac{g(\zeta)}{\zeta - z} d\zeta + \int_{\gamma_2(\epsilon)} \frac{g(\zeta)}{\zeta - z} d\zeta.$$

If one uses the number $w = R^2/\bar{z}$ as in the proof of Poisson's formula in Chapter VIII we get

$$0 = \frac{1}{2\pi i} \int_{\gamma(\epsilon)} \frac{g(\zeta)}{\zeta - w} d\zeta.$$

Subtracting as in Poisson's formula, we find

$$g(z) = \int_{|\theta - \theta_0| \geq \delta} g(Re^{i\theta}) \operatorname{Re} \frac{Re^{i\theta} + z}{Re^{i\theta} - z} \frac{d\theta}{2\pi} + \frac{1}{2\pi i} \int_{\gamma_2(\epsilon)} \left[\frac{g(\zeta)}{\zeta - z} - \frac{g(\zeta)}{\zeta - w} \right] d\zeta$$

$$= g_{1,\epsilon}(z) + g_{2,\epsilon}(z).$$

The length of the curve $\gamma_2(\epsilon)$ is $O(\epsilon)$ for $\epsilon \to 0$. We also have the bound

$$|g(\zeta)| = O(|\log \epsilon|) \quad \text{for } \zeta \text{ on } \gamma_2(\epsilon).$$

Hence the second integral on the right tends to 0 as $\epsilon \to 0$. Thus we may write

$$g(z) = \int_0^{2\pi} \log(Re^{i\theta} - z_0) \operatorname{Re} \frac{Re^{i\theta} + z}{Re^{i\theta} - z} \frac{d\theta}{2\pi}.$$

We can now argue like in Theorem 4.2 of Chapter VIII. By using the fact that $\log(Re^{i\theta} - z_0)$ is absolutely integrable, we may differentiate the function

$$g_1(z) = \int_0^{2\pi} \log|Re^{i\theta} - z_0| \frac{Re^{i\theta} + z}{Re^{i\theta} - z} \frac{d\theta}{2\pi}$$

under the integral sign to see that g_1 is holomorphic on D_R. Since g and g_1 have the same real part it follows that $g - g_1$ is a pure imaginary constant, which proves the desired formula.

Exercise XII.1.5. *Let f be meromorphic. Define $n_f^+(0) = \max(0, n_f(0))$, and:*

$$N_f(\infty, R) = \sum_{a \in D_R, f(a) = \infty, a \neq 0} -(\mathrm{ord}_a\, f) \log \frac{R}{|a|} + n_{1/f}^+(0) \log R,$$

$$N_f(0, R) = \sum_{a \in D_R, f(a) = 0, a \neq 0} (\mathrm{ord}_a\, f) \log \frac{R}{|a|} + n_f^+(0) \log R.$$

Show that Jensen's formula can be written in the form

$$\int_0^{2\pi} \log|f(Re^{i\theta})| \frac{d\theta}{2\pi} + N_f(\infty, R) - N_f(0, R) = \log|c_f|.$$

Solution. After looking at Jensen's formula, we see that it is sufficient to prove that

$$N_f(\infty, R) - N_f(0, R) = \sum_{a \in D_R, a \neq 0} n_f(a) \log|a|/R + n_f(0) \log \frac{1}{R}.$$

Separating zeros and poles we can write

$$\sum_{a \in D_R, a \neq 0} n_f(a) \log|a|/R = \sum_{a \in D_R, f(a) = \infty, a \neq 0} -n_f(a) \log R/|a|$$
$$- \sum_{a \in D_R, f(a) = 0, a \neq 0} n_f(a) \log R/|a|$$

and therefore, we will be done if we can show that

$$n_{1/f}^+(0) \log R - n_f^+(0) \log R = n_f(0) \log 1/R.$$

But this follows at once from considering the cases were the origin is a zero, a pole, or neither.

Exercise XII.1.6. *Let α be a positive real number. Define*

$$\log^+(\alpha) = \max(0, \log \alpha).$$

(a) Show that $\log \alpha = \log^+(\alpha) - \log^+(1/\alpha)$.
(b) Let f be meromorphic on $\overline{D}_R$. For $r < R$ define

$$m_f(r) = \int_0^{2\pi} \log^+ |f(re^{i\theta})| \frac{d\theta}{2\pi} \quad and \quad G_{R,f}^{\infty}(z) = \prod_{a \in D_R, f(a) = \infty} G_R(z, a)^{-\mathrm{ord}_a\, f}.$$

Let $G = G_{R,f}^\infty$. Show that

$$m_G(r) = N_f(\infty, R) - N_f(\infty, r).$$

(c) Following Nevanlinna, define the **height function**

$$T_f(r) = m_f(r) + N_f(\infty, r).$$

Deduce Nevanlinna's formulation of Jensen's formula:

$$T_{1/f}(r) = T_f(r) - \log |c_f|.$$

Solution. (a) If $\alpha \geq 1$, then $1/\alpha \leq 1$ so

$$\log^+(\alpha) - \log^+(1/\alpha) = \log \alpha + 0.$$

If $\alpha < 1$, then $1/\alpha > 1$, so

$$\log^+(\alpha) - \log^+(1/\alpha) = 0 - \log 1/\alpha = \log \alpha.$$

(b) We begin with $m_G(r)$ and work our way to $N_f(\infty, R) - N_f(\infty, r)$. Since $1/G_{r,a}(z)$ is holomorphic on the open disc $D(0, r)$ and bounded by one on the circle, we conclude that $|G_{r,a}(z)| \geq 1$ and we may replace $\log^+$ by $\log$. This gives us, together with Jensen's formula (Exercise 5) and the definitions

$$m_G(r) = \int_0^{2\pi} \log^+ |G(re^{i\theta})| \frac{d\theta}{2\pi}$$

$$= \sum_{\substack{a \in D_R \\ f(a) = \infty}} -n_f(a) \int_0^{2\pi} \log^+ |G_{R,a}(re^{i\theta})| \frac{d\theta}{2\pi}$$

$$= \sum_{\substack{a \in D_R \\ f(a) = \infty}} -n_f(a) \left[\log |c_{G_{R,a}}| - N_{G_{R,a}}(\infty, r) + N_{G_{R,a}}(0, r) \right].$$

In the above, the term $a = 0$ contributes only if 0 is a pole of f and is

$$= -n_f(0) \left[\log R - n_{1/G_{R,0}}^+(0) \log r + n_{G_{R,0}}^+(0) \log r \right]$$

$$= -n_f(0) \left[\log R - \log r \right] = -n_{1/f}^+ \left[\log r - \log R \right].$$

For the terms with $a \neq 0$, we obtain the sums

$$\sum_{\substack{a \in D_R \cap D_r \\ f(a) = \infty \\ a \neq 0}} -n_f(a) \left[\log \frac{R}{|a|} - \log \frac{r}{|a|} \right]$$

$$+ \sum_{\substack{a \in D_R \cap D_r^c \\ f(a) = \infty \\ a \neq 0}} -n_f(a) \left[\log \frac{R}{|a|} \right]$$

$$= \sum_{\substack{a \in D_R \\ f(a) = \infty \\ a \neq 0}} -n_f(a) \log \frac{R}{|a|} - \sum_{\substack{a \in D_r \\ f(a) = \infty \\ a \neq 0}} -n_f(a) \log \frac{r}{|a|}$$

$$= N_f(\infty, R) - n_{1/f}^+(0) \log R - N_f(\infty, r) + n_{1/f}^+(0) \log r$$

$$= N_f(\infty, R) - N_f(\infty, r) + n_{1/f}^+(0) \left[\log r - \log R \right]$$

Adding the term $a = 0$ and the terms $a \neq 0$ concludes the proof.

(c) By definition we see that the difference $T_f(r) - T_{1/f}(r)$ is equal to

$$\int_0^{2\pi} \left(\log^+ |f(re^{i\theta})| - \log^+ |1/f(re^{i\theta})|\right) \frac{d\theta}{2\pi} + N_f(\infty, r) - N_{1/f}(\infty, r).$$

The result in (a) together with the fact that $N_{1/f}(\infty, r) = N_f(0, r)$ gives

$$T_f(r) - T_{1/f}(r) = \int_0^{2\pi} \log |f(re^{i\theta})| \frac{d\theta}{2\pi} + N_f(\infty, r) - N_f(0, r).$$

The preceding exercise implies that

$$T_f(r) - T_{1/f}(r) = \log |c_f|$$

as was to be shown.

XII.2   The Picard–Borel Theorem

Exercise XII.2.1. *Let h be an entire function without zeros. Show that m_h is continuous. [This is essentially trivial, by the uniform continuity of continuous functions on compact sets.]*

Solution. By definition

$$m_h(r) = \int_0^{2\pi} \log^+ |h(re^{i\theta})| \frac{d\theta}{2\pi}.$$

The function $\log^+ |h(re^{i\theta})|$ of two variables r and θ is continuous, so if we fix $t > 0$ and $\epsilon > 0$, we see that for each $0 \leq \theta \leq 2\pi$, there exists a a ball B_θ centered at (t, θ) and of radius s_θ such that for all $(r, \phi) \in B_\theta$ we have

$$\left|\log^+ |h(te^{i\theta})| - \log^+ |h(re^{i\phi})|\right| < \epsilon,$$

and in particular if $\phi = \theta$ we get

$$\left|\log^+ |h(te^{i\theta})| - \log^+ |h(re^{i\theta})|\right| < \epsilon.$$

Since $\bigcup_\theta B_\theta$ is an open cover of the compact interval $0 \leq \theta \leq 2\pi$, we can select a subcover and in particular a number $\delta > 0$ such that if $|t - r| < \delta$, then

$$\left|\log^+ |h(te^{i\theta})| - \log^+ |h(re^{i\theta})|\right| < \epsilon$$

for all $0 \leq \theta \leq 2\pi$. Whence

$$|m_h(t) - m_h(r)| < \epsilon$$

whenever $|t - r| < \delta$, and this proves that m_h is a continuous function.

Exercise XII.2.2. *Let f be a meromorphic function. Show that m_f is continuous. This is less trivial, but still easy. Let $z_1, \ldots, z_s$ be the poles of f on a circle of radius r. Let $z_j = re^{i\theta_j}$, and let $I(\theta_j, \delta)$ be the open interval of radius δ centered* at θ_j. Let I be the union of these intervals. Then for $\theta \notin I$, $f(te^{i\theta})$ converges uniformly to $f(re^{i\theta})$ as $t \to r$, so as $t \to r$,

$$\int_{\theta \notin I} \log^+ |f(te^{i\theta})| \frac{d\theta}{2\pi} \to \int_{\theta \notin I} \log^+ |f(re^{i\theta})| \frac{d\theta}{2\pi}.$$

Given ϵ, there exists δ such that if $\theta \in I(\theta_j, \delta)$ for some j and $|t - r| < \delta$, then

$$\log^+ |f(te^{i\theta})| = \log |f(te^{i\theta})| \quad and \quad \log^+ |f(re^{i\theta})| = \log^+ |f(re^{i\theta})|$$

because $|f(z)|$ is large when z is near z_j. But for z near z_j,

$$f(z) = (z - z_j)^{-e} g(z),$$

where $e > 0$ and $|g(z)|$ is bounded away from 0. Hence for z near z_j,

$$\log |f(z)| = -e \log |z - z_j| + bounded\ function.$$

Then

$$\int_{\theta \in I(\theta_j, \delta)} \log |f(re^{i\theta})| \frac{d\theta}{2\pi} \quad and \quad \int_{\theta \in I(\theta_j, \delta)} \log |f(te^{i\theta})| \frac{d\theta}{2\pi}$$

are both small, because they essentially amount to

$$\int_{-\delta}^{\delta} \log |\theta| d\theta,$$

up to a bounded factor. Put in the details of this part of the argument.]

Solution. Consider the integral

$$J = \int_{\theta \in I(\theta_j, \delta)} \log |f(re^{i\theta})| \frac{d\theta}{2\pi}.$$

We know that for z near z_j we have

$$\log |f(z)| = -e \log |z - z_j| + bounded\ function.$$

The integral of a bounded function on $I(\theta_j, \delta)$ will be $\leq 2\delta C$ for some constant C, so we are done if we can show that the integral

$$\int_{\theta \in I(\theta_j, \delta)} \log |re^{i\theta} - z_j| \frac{d\theta}{2\pi}.$$

is small. We have $z_j = re^{i\theta_j}$, so changing variables $\varphi = \theta - \theta_j$ in the above integral we find

$$\int_{-\delta}^{\delta} \log \left(r|e^{i\varphi} - 1| \right) \frac{d\varphi}{2\pi} = \int_{-\delta}^{\delta} \log r \frac{d\varphi}{2\pi} + \int_{-\delta}^{\delta} \log |e^{i\varphi} - 1| \frac{d\varphi}{2\pi}.$$

The last integral on the right is essentially $\int_{-\delta}^{\delta} \log |\varphi| d\varphi$ because for all φ near 0 we have

$$\frac{1}{2} |\varphi| \leq |e^{i\varphi} - 1| \leq |\varphi|.$$

To see this, draw a picture, or square each side, substitute sine and cosine and use elementary calculus.

For the integral

$$\int_{\theta \in I(\theta_j, \delta)} \log |f(te^{i\theta})| \frac{d\theta}{2\pi}$$

we are reduced to estimating the integral

$$\int_{\theta \in I(\theta_j, \delta)} \log |te^{i\theta} - z_j| \frac{d\theta}{2\pi}.$$

Changing variables $\varphi = \theta - \theta_j$ we find that this last integral is equal to

$$\int_{-\delta}^{\delta} \log r \frac{d\varphi}{2\pi} + \int_{-\delta}^{\delta} \log \left| e^{i\varphi} - \frac{t}{r} \right| \frac{d\varphi}{2\pi}.$$

Then for all t/r close to 1 and all φ near 0

$$c|\varphi| \le \left| e^{i\varphi} - \frac{t}{r} \right|$$

where c is a small positive constant. This concludes the exercise.

Exercise XII.2.3. *Let f be meromorphic. Let a be a complex number. Show that*

$$T_f(r) = T_{f-a}(r) + O_a(1),$$

where $|O_a(1)| \le \log^+ |a| + \log 2$.

Solution. We use the notation, assumptions, and results of Exercise 5 and 6 of the preceding section. Jensen's formula applied to $f - a$ gives

$$\log |c_{f-a}| + N_{f-a}(0, r) = \int_0^{2\pi} \log |f(re^{i\theta}) - a| \frac{d\theta}{2\pi} + N_{f-a}(\infty, r).$$

From the definition we have $N_{f-a}(r, \infty) = N_f(r, \infty)$. We also know that $\log \alpha = \log^+ \alpha - \log^+ 1/\alpha$ so letting $\alpha = |f(re^{i\theta}) - a|$ and using the definition of m_{f-a} we obtain

$$\log |c_{f-a}| + N_{f-a}(0, r) + m_{1/(f-a)}(r) = \int_0^{2\pi} \log^+ |f(re^{i\theta}) - a| \frac{d\theta}{2\pi} + N_f(\infty, r).$$

The left hand side of the above equality can be rewritten as $\log |c_{f-a}| + T_{1/(f-a)}$ because $N_{f-a}(0, r) = N_{1/(f-a)}(\infty, r)$. Applying Exercise 6 of §1, we get

$$T_{f-a}(r) = \int_0^{2\pi} \log^+ |f(re^{i\theta}) - a| \frac{d\theta}{2\pi} + N_f(\infty, r).$$

The triangle inequality implies

$$\log^+ |f(re^{i\theta}) - a| \le \log^+(|f(re^{i\theta})| + |a|).$$

However, if $\alpha, \beta > 0$ we have

$$\log^+(\alpha + \beta) \le \log^+ \alpha + \log^+ \beta + \log 2$$

and this inequality follows from looking at $\log^+(2 \max(\alpha, \beta))$. Using the triangle inequality once more, we find

$$|f(re^{i\theta}) - a| \le |f(re^{i\theta})| + |a|$$

and

$$|f(re^{i\theta})| \le |f(re^{i\theta}) - a| + |a|.$$

The previous remark implies

$$\log^+ |f(re^{i\theta}) - a| \le \log^+ |f(re^{i\theta})| + \log^+ |a| + \log 2$$

and

$$\log^+ |f(re^{i\theta})| \le \log^+ |f(re^{i\theta}) - a| + \log^+ |a| + \log 2,$$

whence

$$T_f(r) - \log^+ |a| - \log 2 \le T_{f-a}(r) \le T_f(r) + \log^+ |a| + \log 2$$

as was to be shown.

We give another proof not using Jensen's inequality. By definition

$$T_{f-a}(r) = \int_0^{2\pi} \log^+ |f(re^{i\theta}) - a| \frac{d\theta}{2\pi} + N_{f-a}(\infty, R).$$

But from the definitions, it is clear that $N_{f-a}(\infty, R) = N_f(\infty, R)$ so

$$T_{f-a}(r) - T_f(r) = \int_0^{2\pi} \log^+ |f(re^{i\theta}) - a| \frac{d\theta}{2\pi} - \int_0^{2\pi} \log^+ |f(re^{i\theta})| \frac{d\theta}{2\pi}$$

$$= \int_0^{2\pi} \left(\log^+ |f(re^{i\theta}) - a| - \log^+ |f(re^{i\theta})| \right) \frac{d\theta}{2\pi}.$$

Now conclude using the inequalities we used in the first proof.

XII.6   The Phragmen–Lindelöf and Hadamard Theorems

Exercise XII.6.1. *Let U be the right half plane (Re(z) > 0). Let f be continuous on the closure of U and analytic on U. Assume that there are constants C > 0 and α < 1 such that*

$$|f(z)| \le Ce^{|z|^\alpha}$$

for all z in U. Assume that f is bounded by 1 on the imaginary axis. Prove that f is bounded by 1 on U. Show that this assertion is not true if α = 1.

Solution. If $\alpha = 1$ the assertion is not true, because the function $f(z) = e^z$ gives a counter example.

In the case $\alpha < 1$, we give two proofs. In the first proof we reduce the problem to the Phragmen–Lindelöf theorem. If S denotes the strip

$$S = \{x + iy : -\frac{\pi}{2} < x < \frac{\pi}{2}\},$$

consider the isomorphism $\varphi : S \to$ right half plane, defined by

$$\varphi(w) = e^{iw}.$$

We now show that the function $h = f \circ \varphi$ satisfies the hypothesis of the Phragmen–Lindelöf theorem. The function φ maps the boundary of S onto the boundary of the right half plane, so $|h(w)| \leq 1$ on the sides of the strip and we have

$$|h(w)| \leq Ce^{|\varphi(w)|^\alpha} \leq Ce^{e^{\alpha|w|}}.$$

We can apply the Phragmen–Lindelöf theorem to conclude that $|h(w)| \leq 1$ on the strip. Hence $|f(z)| \leq 1$ in the right half plane as was to be shown.

We now give a direct proof which is modeled on the proof of the Phragmen–Lindelöf theorem. Choose a number γ such that $\alpha < \gamma < 1$. Let $z_0 = -1$ and take the branch of $\log(z - z_0)$ obtained by deleting $(-\infty, z_0]$ from the real axis and taking the angles from $-\pi$ to π. For each $\epsilon > 0$, define a function g_ϵ is the right half plane by

$$g_\epsilon(z) = f(z)e^{-\epsilon(z-z_0)^\gamma}.$$

This function is analytic on the right half plane and extends continuously on the imaginary axis. Then

$$|g_\epsilon(z)| = |f(z)|e^{-\epsilon|z-z_0|^\gamma \cos\gamma\theta}$$

where $\theta = \arg(z - z_0)$. For z in the right half plane we have $-\pi/2 \leq \theta \leq \pi/2$ hence $-\gamma\pi/2 \leq \gamma\theta \leq \gamma\pi/2$ and this guarantees the existence of a constant $c > 0$ such that $\cos\gamma\theta > c$ for all z in the right half plane. Hence

$$|g_\epsilon(z)| \leq |f(z)|e^{-\epsilon|z-z_0|^\gamma c},$$

and for all large $|z|$ we get

$$|g_\epsilon(z)| \leq Ce^{|z|^\alpha}e^{-\epsilon|z-z_0|^\gamma c}.$$

This estimate implies that for z in the right half plane, we have the limit

$$\lim_{|z|\to\infty} |g_\epsilon(z)| = 0.$$

On the imaginary axis we also have $|g_\epsilon(z)| \leq |f(z)| \leq 1$. Let $\overline{D}_{\text{right}}(R)$ denote the closed right half disc centered at the origin and of radius R. The above results together with the maximum modulus principle applied to $\overline{D}_{\text{right}}(R)$ with arbitrarily large R shows that $|g_\epsilon(z)| \leq 1$ in the right half plane. Now, if R is fixed, we have $|g_\epsilon(z)| \leq 1$ in $\overline{D}_{\text{right}}(R)$ so

$$|f(z)| \leq e^{\epsilon|z-z_0|^\gamma \cos\gamma\theta} \leq e^{\epsilon(R+|z_0|)^\gamma}$$

and letting $\epsilon \to 0$ we obtain $|f(z)| \leq 1$ in $\overline{D}_{\text{right}}(R)$. Since R was arbitrary, we conclude that $|f(z)| \leq 1$ in the right half plane, as was to be shown.

Exercise XII.6.2. *More generally, let U be the open sector between two rays from the origin. Let f be continuous on the closure of U (i.e., the sector and rays), and analytic on U. Assume that there are constants C > 0 and α such that*

$$|f(z)| \le Ce^{|z|^\alpha}$$

for all z in U. If π/β is the angle of the sector, assume that $0 < \alpha < \beta$. If f is bounded by 1 on the rays, prove that f is bounded by 1 on U.

Solution. Let H be the right half plane. We show how this exercise reduces to Exercise 1. There exists $\theta \in \mathbf{R}$ such that $\psi : U \to H$ defined by

$$\psi(z) = e^{-i\pi/2}(e^{i\theta}z)^\beta$$

is an isomorphism. This result comes from the following sequence of transformations

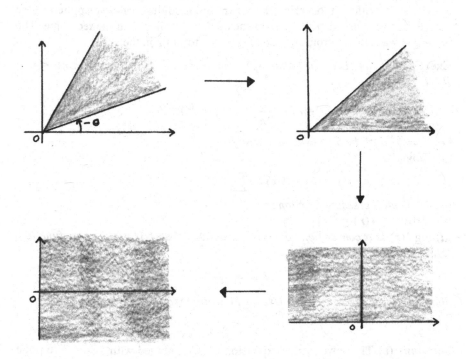

Let $g = f \circ \psi^{-1}$. From reversing the above steps we find

$$|\psi^{-1}(s)| = |s|^{1/\beta},$$

thus

$$|g(s)| \le Ce^{|s|^{\alpha/\beta}}$$

with $\alpha/\beta < 1$. Clearly, $|g(s)| \leq 1$ on the imaginary axis because $|f(z)| \leq 1$ on the rays of the sector U. Exercise 1 applies to this situation, so $|g(s)| \leq 1$ on the right half plane. Hence $|f(s)| \leq 1$ in the sector.

Exercise XII.6.3. *Consider again a finite strip $\sigma_1 \leq \sigma \leq \sigma_2$. Suppose that f is holomorphic on the strip, $|f(s)| \to 0$ as $|s| \to \infty$ with s in the strip, and $|f(s)| \leq 1$ on the sides of the strip. Prove that $|f(s)| \leq 1$ in the strip.*

Solution. We assume that f extends to a continuous function on the boundary of the strip. This exercise is a simple application of the maximum modulus principle. Take a truncated rectangle in the strip, say

$$R(T) = \{x + iy : \sigma_1 \leq x \leq \sigma_2 \quad \text{and} \quad -T \leq y \leq T\},$$

and select a disc centered at 0 of radius so large that the rectangle is contained in the disc, and such that for all z in the strip and z on the boundary of the disc we have $|f(z)| \leq 1$. This can be done because of the assumption that $|f(s)| \to 0$ as $|s| \to \infty$ with s in the strip. The maximum modulus principle applied to the closed disc intersected with the strip shows that $|f(z)| \leq 1$ in the rectangle. The rectangle was chosen arbitrarily, so $|f(z)| \leq 1$ for all z in the strip.

Exercise XII.6.4. *Let f be holomorphic on the disc D_R of radius R. For $0 \leq r < R$ let*

$$I(r) = \frac{1}{2\pi} \int_0^{2\pi} |f(re^{i\theta})|^2 d\theta.$$

Let $f = \sum a_n z^n$ be the power series of f.
(a) Show that

$$I(r) = \sum |a_n|^2 r^{2n}.$$

(b) I(r) is an increasing function of r.
(c) $|f(0)|^2 \leq I(r) \leq \|f\|_r^2$.
(d) $\log I(r)$ is a convex function of $\log r$, assume that f is not the zero function.
[Hint: Put $s = \log r$,

$$J(s) = I(e^s).$$

Show that $(\log J)'' = \frac{J''J - (J^2)}{J^2}$. Use the Schwarz inequality to show that

$$J''J - (J')^2 \geq 0.]$$

Solution. (a) The power series expansion of f combined with a switch to polar coordinates give the following expressions

$$f(re^{i\theta}) = \sum a_n r^n e^{ni\theta} \quad \text{and} \quad \overline{f(re^{i\theta})} = \sum \overline{a}_n r^n e^{-ni\theta}.$$

When we integrate from 0 to 2π the function $e^{ik\theta}$ where $k \in \mathbf{Z}$ and $k \neq 0$, we get 0, so multiplying the above two series we obtain

$$I(r) = \frac{1}{2\pi} \int_0^{2\pi} \sum |a_n|^2 r^{2n} d\theta = \sum |a_n|^2 r^{2n},$$

as was to be shown.

(b) For each n the function $r \mapsto r^{2n}$ is increasing, so the formula of (a) implies that $I(r)$ is an increasing function of r.

(c) The first inequality is a consequence of (a), and the second inequality can be proved as follows,

$$\frac{1}{2\pi} \int_0^{2\pi} |f(re^{i\theta})| d\theta \le \frac{1}{2\pi} \int_0^{2\pi} \|f\|_r^2 d\theta \le \|f\|_r^2.$$

(d) The basic rules of differentiation imply

$$(\log J)'' = \left(\frac{J'}{J}\right)' = \frac{J''J - (J')^2}{J^2}.$$

Since $r = e^s$ we have

$$J = \sum |a_n|^2 e^{2sn}, \quad J' = \sum |a_n|^2 (2n) e^{2sn} \quad \text{and} \quad J' = \sum |a_n|^2 (4n^2) e^{2sn}.$$

In ℓ^2 (the space of complex sequences $\{x_n\}$ such that $\sum |x_n|^2$ converges) we have the hermitian product

$$\langle \{u_n\}, \{v_n\}\rangle = \sum u_n \bar{v}_n.$$

Letting $u_n = a_n(2n)e^{sn}4$ and $v_n = a_n e^{sn}$ we see that after applying the Schwarz inequality

$$\langle \{u_n\}, \{v_n\}\rangle^2 \le \langle \{u_n\}, \{u_n\}\rangle \langle \{v_n\}, \{v_n\}\rangle$$

we get $(J')^2 \le J''J$, as was to be shown.

XIII
Entire and Meromorphic Functions

XIII.1 Infinite Products

Exercise XIII.1.1. *Let $0 < |\alpha| < 1$ and let $|z| \le r < 1$. Prove the inequality*

$$\left| \frac{\alpha + |\alpha|z}{(1 - \overline{\alpha}z)\alpha} \right| \le \frac{1 + r}{1 - r}.$$

Solution. We have

$$\left| \frac{\alpha + |\alpha|z}{(1 - \overline{\alpha}z)\alpha} \right| = \left| \frac{1 + (\frac{|\alpha|}{\alpha})z}{1 - \overline{\alpha}z} \right| \le \frac{1 + r}{1 - r}.$$

Exercise XIII.1.2 (Blaschke Products). *Let $\{\alpha_n\}$ be a sequence in the unit disc D such that $\alpha_n \ne 0$ for all n, and*

$$\sum_{n=1}^{\infty}(1 - |\alpha_n|)$$

converges. Show that the product

$$f(z) = \prod_{n=1}^{\infty} \frac{\alpha_n - z}{1 - \overline{\alpha}_n z} \frac{|\alpha_n|}{\alpha_n}$$

converges uniformly for $|z| \le r < 1$, and defines a holomorphic function on the unit disc having precisely the zeros α_n and no other zeros. Show that $|f(z)| \le 1$.

Solution. Suppose $|z| \leq r < 1$. Let

$$f_n(z) = \frac{\alpha_n - z}{1 - \bar{\alpha}_n z} \frac{|\alpha_n|}{\alpha_n},$$

and $h_n(z) = 1 - f_n(z)$. Then a direct computation shows that

$$|h_n(z)| = \left| \frac{\alpha_n + |\alpha_n|z}{(1 - \bar{\alpha}_n z)\alpha_n} \right| |1 - |\alpha_n||,$$

so by Exercise 1 we get the bound

$$|h_n(z)| \leq C(1 - |\alpha_n|)$$

where $C = (1+r)/(1-r)$. For all large n we have $|h_n(z)| \leq 1/2$ and by definition $f_n = 1 - h_n$ so there exists a positive constant K such that for all large n we get

$$|\log f_n(z)| \leq K|h_n(z)| \leq KC(1 - |\alpha_n|).$$

Since the series $\sum(1 - |\alpha_n|)$ converges, we conclude that the series $\sum \log f_n(z)$ converges uniformly and so does the product $\prod f_n(z)$. Hence the product defines a holomorphic function on the unit disc.

Each term in the product is ≤ 1 so $|f(z)| \leq 1$. We now show that f has the desired zeros, and to do this we argue as in § 2 of Lang's book. We fix some radius r and consider on $|z| \leq r$. From the previous inequalities and the fact that $\sum(1 - |\alpha_n|)$ converges, we see that given ϵ there exists N_0 such that if $N \geq N_0$ then

$$\left| \log \prod_{n=N_0}^{N} f_n(z) \right| < \epsilon.$$

So the product $\prod_{n=N_0}^{N} f_n(z)$ is close to 1. By definition

$$f(z) = \prod_{n=1}^{N_0-1} f_n(z) \lim_{N \to \infty} \prod_{n=N_0}^{N} f_n(z).$$

The first product on the right has the appropriate zeros in the disc $|z| \leq r$. The limit of the second product on the right is close to 1, and hence has no zeros.

Exercise XIII.1.3. *Let $\alpha_n = 1 - 1/n^2$ in the preceding exercise. Prove that*

$$\lim_{x \to 1} f(x) = 0 \quad \text{if } 0 < x < 1.$$

In fact, prove the estimate for $\alpha_{n-1} < x < \alpha_n$:

$$|f(x)| < \prod_{k=1}^{n-1} \frac{x - \alpha_k}{1 - \alpha_k x} < \prod_{k=1}^{n-1} \frac{\alpha_n - \alpha_k}{1 - \alpha_k} < 2e^{-n/3}.$$

Solution. We have $\sum 1 - |\alpha_n| = \sum 1/n^2$ so the Blaschke product converges. Suppose that $\alpha_{n-1} < x < \alpha_n$ and that $k \geq n$. Then $x < \alpha_k$ and $\alpha_k x < 1$ so

$$0 < \frac{\alpha_k - x}{1 - \alpha_k x}.$$

We also have

$$\alpha_k - x = 1 - \frac{1}{k_2} - x < 1 - x + \frac{x}{k^2} = 1 - \alpha_k x$$

thus

$$0 < \frac{\alpha_k - x}{1 - \alpha_k x} < 1$$

and therefore

$$\prod_{k=n}^{\infty} \left| \frac{\alpha_k - x}{1 - \alpha_k x} \right| < 1$$

because from Exercise 2 we know that the product converges. This proves the first inequality

$$|f(x)| < \prod_{k=1}^{n-1} \frac{x - \alpha_k}{1 - \alpha_k x}.$$

If $k < n$, then $x - \alpha_k < \alpha_n - \alpha_k$ and $1 - x\alpha_k > 1 - \alpha_k$ so we have the second inequality, namely

$$\prod_{k=1}^{n-1} \frac{x - \alpha_k}{1 - \alpha_k x} < \prod_{k=1}^{n-1} \frac{\alpha_n - \alpha_k}{1 - \alpha_k}.$$

To establish the third inequality we first note that

$$\frac{\alpha_n - \alpha_k}{1 - \alpha_k} = 1 - \left(\frac{k}{n} \right)^2$$

and since $\log(1 - x) \leq -x$ we get

$$\sum_{k=1}^{n-1} \log \left(\frac{\alpha_n - \alpha_k}{1 - \alpha_k} \right) \leq \sum_{k=1}^{n-1} -\left(\frac{k}{n} \right)^2 = -\frac{1}{n^2} \frac{(n-1)n(2n-1)}{6}.$$

But $(n - 1)n(2n - 1) \geq 2n^3 - 3n^2$ so

$$\sum_{k=1}^{n-1} \log \left(\frac{\alpha_n - \alpha_k}{1 - \alpha_k} \right) \leq -\frac{n}{3} + \frac{1}{2} < -\frac{n}{3} + \log 2.$$

Exponentiating we see that the last inequality

$$\prod_{k=1}^{n-1} \frac{\alpha_n - \alpha_k}{1 - \alpha_k} < 2e^{-n/3}$$

drops out. It is now clear that $\lim_{x \to 1} f(x) = 0$.

Exercise XIII.1.4. *Prove that there exists a bounded analytic function f on the unit disc for which each point of the unit circle is a singularity.*

Solution. We use the previous exercise and the fact that $\{e^{2\pi i r}\}_{r \in \mathbf{Q}}$ is dense on the unit circle. Define a sequence $\{\alpha_{n,k}\}_{1 < n,\, 1 \le k \le n}$ in the unit disc by

$$\alpha_{n,k} = \left(1 - \frac{1}{n^3}\right) e^{2\pi i \frac{k}{n}}.$$

Then $|\alpha_{n,k}| = 1 - 1/n^3$ and therefore

$$\sum (1 - |\alpha_{n,k}|) = \sum_{1<n}\sum_{1 \le k \le n}(1 - |\alpha_{n,k}|) = \sum_{1<n}\frac{1}{n^2} < \infty.$$

Exercise 2 implies that the product

$$f(z) = \prod_{1<n, 1 \le k \le n} \frac{\alpha_{n,k} - z}{1 - \overline{\alpha}_{n,k}z}\frac{|\alpha_{n,k}|}{\alpha_{n,k}}$$

defines a holomorphic function on the unit disc which satisfies the desired properties.

Exercise XIII.1.5 (*q*-Products). *Let $z = x + iy$ be a complex variable, and let $\tau = u + iv$ with u, v real, $v > 0$ be a variable in the upper half-plane H. We define*

$$q_\tau = e^{2\pi i \tau} \quad \text{and} \quad q_z = e^{2\pi i z}.$$

Consider the infinite product

$$(1 - q_z)\prod_{n=1}^{\infty}(1 - q_\tau^n q_z)(1 - q_\tau^n/q_z).$$

(a) Prove that the infinite product is absolutely convergent.
(b) Prove that for fixed τ, the infinite product defines a holomorphic function of z, with zeros at the points

$$m + n\tau, \quad m, n \text{ integers.}$$

*We define the **second Bernoulli polynomial***

$$\mathbf{B}_2(y) = y^2 - y + \frac{1}{6}.$$

*Define the **Néron–Green function***

$$\lambda(z, \tau) = \lambda(x, y, \tau) - \log\left|q_\tau^{\mathbf{B}_2(y/v)/2}(1 - q_z)\prod_{n=1}^{\infty}(1 - q_\tau^n q_z)(1 - q_\tau^n/q_z)\right|.$$

(c) Prove that for fixed τ, the function $z \mapsto \lambda(z, \tau)$ is periodic with periods 1, τ.

Solution. (a) We estimate the partial products separately using the criterion of Lemma 1.1. We get

$$|q_\tau^n q_z| = e^{-2\pi n v}e^{-2\pi y} \quad \text{and} \quad |q_\tau^n q_z^{-1}| = e^{-2\pi n v}e^{2\pi y}.$$

The series $\sum e^{-2\pi n v}$ converges because $v > 0$ so the product converges absolutely.

(b) Let τ be fixed and let $R > 0$. Suppose $|z| \leq R$. Then from the previous estimates we see that we have

$$|q_\tau^n q_z| = e^{-2\pi n v} e^{-2\pi R} \quad \text{and} \quad |q_\tau^n q_z^{-1}| = e^{-2\pi n v} e^{2\pi R}$$

so the product converges uniformly on $|z| \leq R$. This is true for all R, so the product defines a holomorphic function of z. Now for $|z| < R$ we can write the product $\prod (1 - q_\tau^n q_z)(1 - q_\tau^n / q_z)$ as

$$\prod_{n=1}^{N}(1 - q_\tau^n q_z)(1 - q_\tau^n / q_z) \prod_{n=N+1}^{\infty} (1 - q_\tau^n q_z)(1 - q_\tau^n / q_z)$$

where N is so large that the absolute value of the product on the right is close to 1 for all $|z| < R$. Then the zeros of the function on $|z| < R$ are determined by $(1 - q_z) \prod_{n=1}^{N}(1 - q_\tau^n q_z)(1 - q_\tau^n / q_z)$. We analyze the zeros of the three factors in parenthesis. For the first factor we have $q_z = 1$ if and only if $z \in \mathbf{Z}$. For the second factor we have

$$q_\tau^n q_z = e^{2\pi i(n\tau + z)} = 1$$

if and only $n\tau + z \in \mathbf{Z}$. Finally for the third factor we have $q_\tau^n q_{-z} = 1$ if and only if $n\tau - z \in \mathbf{Z}$. The argument holds for any $R > 0$ so the product defines a holomorphic function whose zeros are $m + n\tau$ where m and n are integers.

(c) Since $q_{z+1} = e^{2\pi i(z+1)} = e^{2\pi i z} = q_z$, it is immediate that

$$\lambda(z + 1, \tau) = \lambda(z, \tau).$$

We now show that τ is also a period. The expression in the absolute value of $\lambda(z + \tau, \tau)$ is

$$q_\tau^{\mathbf{B}_2((y+v)/v)/2}(1 - q_{z+\tau}) \prod_{n=1}^{\infty}(1 - q_\tau^n q_{z+\tau})(1 - q_\tau^n q_{z+\tau}^{-1}).$$

But $q_\tau^n q_{z+\tau} = q_\tau^{n+1} q_z$, $q_\tau^n q_{z+\tau}^{-1} = q_\tau^{n-1} q_z$ and $\mathbf{B}_2((y + v)/v) = \mathbf{B}_2(y/v) + 2y/v$ so multiplying by the appropriate terms so as to keep the product unchanged, we see that the above expression is equal to

$$q_\tau^{y/v} q_\tau^{\mathbf{B}_2(y/v)/2}(1 - q_{z+\tau}) \left(\frac{1}{1 - q_\tau q_z} \right) (1 - q_z^{-1}) \prod_{n=1}^{\infty}(1 - q_\tau^n q_z)(1 - q_\tau^n q_z^{-1})$$

which after some simplifications becomes

$$-\frac{q_\tau^{y/v}}{q_z} q_\tau^{\mathbf{B}_2(y/v)/2}(1 - q_z) \prod_{n=1}^{\infty}(1 - q_\tau^n q_z)(1 - q_\tau^n q_z^{-1}).$$

The periodicity of λ follows from the fact that

$$\left| -\frac{q_\tau^{y/v}}{q_z} \right| = \frac{e^{-2\pi y}}{e^{-2\pi y}} = 1.$$

XIII.2 Weierstrass Products

Exercise XIII.2.1. *Let f be an entire function and n a positive integer. Show that there is an entire function such that $g^n = f$ if and only if the orders of the zeros of f are divisible by n.*

Solution. Suppose that there exists an entire function g such that $g^n = f$. Then z is a zero for g if and only if z is a zero of f, and from the power series expansions at z is is clear that $\text{ord}_z(g^n) = n \cdot \text{ord}_z(g)$. But $\text{ord}_z(g^n) = \text{ord}_z(f)$, so the orders of the zeros of f are divisible by n.

Conversely, suppose that the orders of the zeros of f are divisible by n. Let $z_1, z_2, \ldots$, be the zeros of f and $r_i = (\text{ord}_{z_i} f)/n$. Let h be an entire function with zeros $\{z_i\}$ of orders r_i respectively. Then f/h^n is entire and has no zeros. Let

$$\tilde{h}(z) = e^{\frac{1}{n} \log(f(z)/h^n(z))}.$$

Then $\tilde{h}^n = f/h^n$ so $g = \tilde{h}h$ is entire, and $g^n = f$.

Exercise XIII.2.2. *Prove that*

$$\sum_{n=1}^{\infty} \frac{1}{n^2} = \frac{\pi^2}{6}.$$

[Hint: Use the constant term of the Laurent expansion of $\pi^2/\sin^2 \pi z$ at $z = 0$.]

Solution. We know that

$$\frac{\pi^2}{\sin^2 \pi z} = \sum_{n \in \mathbb{Z}} \frac{1}{(z-n)^2} = \sum_{n \in \mathbb{Z}-\{0\}} \frac{1}{(z-n)^2} + \frac{1}{z^2},$$

so it is sufficient to find the constant term in the Laurent expansion of the left hand side at $z = 0$. Inverting the series of the sine function we get

$$\frac{1}{\sin T} = \frac{1}{T} \frac{1}{1 - T^2/3! + T^5/5! - \cdots}$$
$$= \frac{1}{T} \left(1 + (T^2/3! - T^5/5! + \cdots) + (T^2/3! - T^5/5! + \cdots)^2 + \cdots\right).$$

Squaring and making the substitution $T = \pi z$ we get

$$\frac{1}{\sin^2 \pi z} = \frac{1}{\pi^2 z^2} + \frac{2}{3!} + \text{higher order terms},$$

so

$$\frac{\pi^2}{\sin^2 \pi z} = \frac{1}{z^2} + \frac{2\pi^2}{6} + \text{higher order terms}.$$

Comparing constant terms we get

$$\frac{2\pi^2}{6} = \sum_{n \in \mathbb{Z}-\{0\}} \frac{1}{n^2} = 2\sum_{n=1}^{\infty} \frac{1}{n^2}.$$

Exercise XIII.2.3. *More generally, show:*
(a) $\pi z \cot \pi z = 1 - 2 \sum_{n=1}^{\infty} \sum_{m=1}^{\infty} z^{2m}/n^{2m}$.
(b) *Define the **Bernoulli numbers** B_k by the series*

$$\frac{t}{e^t - 1} = 1 - \frac{t}{2} + \sum_{k=2}^{\infty} B_k \frac{t^k}{k!}.$$

Setting $t = 2i\pi z$ and comparing coefficients, prove:
  If k is an even positive integer, then

$$2\zeta(k) = -\frac{B_k}{k!}(2\pi i)^k.$$

If $k = 2$, you recover the computation of Exercise 2.

Solution. (a) We use the formula given in the text, namely

$$\pi \cot \pi z = \frac{1}{z} + \sum_{n \neq 0}^{\infty} \left(\frac{1}{z - n} + \frac{1}{n} \right).$$

Multiplying by z and splitting the sum over the positive and negative integers we find

$$\pi z \cot \pi z = 1 + \sum_{n=1}^{\infty} \left(\frac{z}{z - n} + \frac{z}{z + n} \right) = 1 + \sum_{n=1}^{\infty} \frac{2z^2}{z^2 - n^2}$$

$$= 1 + \sum_{n=1}^{\infty} \frac{-2z^2}{n^2} \frac{1}{1 - (z^2/n^2)} = 1 - 2 \sum_{n=1}^{\infty} \sum_{m=1}^{\infty} \frac{z^{2m}}{n^{2m}}.$$

(b) In Exercise 1, §1 of Chapter II we proved the formula

$$\pi z \cot \pi z = \sum_{m=0}^{\infty} (-1)^m \frac{(2\pi)^{2m}}{(2m)!} B_{2m} z^{2m}.$$

Interchanging the sum signs in the formula obtained in (a), we see that combined with the above expression of $\pi z \cot \pi z$ we get

$$\frac{(2\pi i)^{2m}}{(2m)!} B_{2m} = -2 \sum_{m=0}^{\infty} \frac{1}{n^{2m}} = -2\zeta(2m),$$

where we used $(-1)^m = (i)^{2m}$. We evaluated the first Bernoulli numbers in Exercise II.1.3 so we get $\zeta(2) = \frac{\pi^2}{6}$ and $\zeta(4) = \frac{\pi^4}{90}$.

Exercise XIII.2.4. *In the terminology of algebra, the set E of entire functions is a ring, and in fact a subring of the ring of all functions; namely E is closed under addition and multiplication, and contains the function 1. By and **ideal** J, we mean a subset of E such that if $f, g \in J$ then $f + g \in J$, and if $h \in E$ then $hf \in J$. In other words, J is closed under multiplication by elements of E, and under addition. If there exists functions $f_1, \ldots, f_r \in J$ such that all elements of J can be expressed in the form $A_1 f_1 + \cdots + A_r f_r$ with $A_i \in E$, then we call $f_1, \ldots, f_r$ **generators** of J, and we say that J is **finitely generated**. Give an example of an ideal of E which is not finitely generated.*

Solution. Let J be the set of functions which are entire and such that $f(n) = 0$ for all but finitely many integers n. Using the Weierstrass product we can construct a function g whose set of zeros is $\mathbb{Z}$, so J is nonempty, and J is an ideal of the ring of entire function E. Suppose that J is finitely generated, say by $f_1, \ldots, f_r$ with $f_i \in J$ for all i. Let Z_i be the set of integer zeros of f_i. Then, the set Z defined by

$$Z = \bigcap_{i=1}^{r} Z_i$$

misses only finitely many integers. Now choose some $m \in Z$. Using Weierstrass products, we can construct a function h whose only zeros are at the points of $Z - \{m\}$. It is now clear that h cannot be written as a sum

$$h = A_1 f_1 + \cdots + A_r f_f$$

because such a sum vanishes at m while h does not. Thus J is not finitely generated. In the terminology of algebra, we have just shown that the ring of entire functions is not Noetherian.

XIII.3   Functions of Finite Order

Exercise XIII.3.1. *Let f, g be entire of order ρ. Show that fg is entire of order $\leq \rho$, and $f + g$ is entire of order $\leq \rho$.*

Solution. Let $\epsilon > 0$. Then for all large R

$$\log \| fg \|_R \leq \log \| f \|_R + \log \| g \|_R \leq C_1 R^{\rho+\epsilon} + C_2 R^{\rho+\epsilon} \leq (C_1 + C_2) R^{\rho+\epsilon},$$

and this proves that fg has order $\leq \rho$. For the sum we see that

$$\| f + g \|_R \leq \| f \|_R + \| g \|_R \leq A^{R^{\rho+\epsilon}} + B^{R^{\rho+\epsilon}} \leq 2C^{R^{\rho+\epsilon}}$$

where $C = \max(A, B)$. Now we may choose $C \geq 2$ so that for all large R

$$\| f + g \|_R \leq D^{R^{\rho+\epsilon}}$$

where $D = C^2$.

Exercise XIII.3.2. *Let f, g be entire of order ρ, and suppose f/g is entire. Show that f/g is entire of order $\leq \rho$.*

Solution. By Hadamard's theorem we have

$$\frac{f(z)}{g(z)} = \frac{e^{h_1(z)} z^{m_1} \prod \left(1 - \frac{z}{z_n^f}\right) e^{P(z/z_n^f)}}{e^{h_2(z)} z^{m_2} \prod \left(1 - \frac{z}{z_n^g}\right) e^{P(z/z_n^g)}}$$

where h_1 and h_2 are polynomials of degree $\leq \rho$. The infinite product in the denominator will cancel because f/g is entire and furthermore $m_1 \geq m_2$. The remaining expression shows that f/g has order $\leq \rho$.

XIII.4   Meromorphic Functions, Mittag–Leffler Theorem

Exercise XIII.4.1. *Let g be a meromorphic function on* **C**, *with poles of order at most one, and integral residues. Show that there exists a meromorphic function f such that $f'/f = g$.*

Solution. Let S_1 be the set of points where g has a pole with positive residue, and S_2 the set of points where g has a pole with negative residue. Using Weierstrass products, we can construct entire functions F and G having the following properties: F has zeros at points of S_1 with order the residue of g at that point, and G has zeros at points of S_2 with order the absolute value of the residue of g at that point. We can now define an entire function h by:

$$h = g - \frac{E'}{E} + \frac{G'}{G}.$$

We also define $\varphi = \exp\left(\int_{z_0}^{z} h\right)$, so that φ is entire, nowhere zero and $\varphi'/\varphi = h$. Finally, we let $f = \varphi E/G$ which is meoromorphic and satisfies

$$\frac{f'}{f} = \frac{\varphi'}{\varphi} + \frac{E'}{E} - \frac{G'}{G} = g.$$

This conludes the exercise.

Exercise XIII.4.2. *Given entire functions f, g without common zeros, prove that there exists entire functions A, B such that $Af + Bg = 1$.* [Hint: *By Mittag–Leffler, there exists a meromorphic function M whose principal parts occur only at the zeros of g, and such that the principal part $\mathrm{Pr}(M, z_n)$ at a zero z_n of g is the same as $\mathrm{Pr}(1/fg, z_n)$, so $M - 1/fg$ is holomorphic at z_n. Let $A = Mg$, and take it from there.*]

Solution. We use the notation of the hint. Let $h = M - 1/fg$. Then h is holomorphic at the zeros of g, but not at the zeros of f. Let $A = Mg$ and $B = -hf$. Then both A and B are entire by construction, and

$$Af + Bg = Mgf - hfg = Mgf - Mfg + 1 = 1$$

as was to be shown.

Exercise XIII.4.3. *Let f, g be entire functions.*
(a) Show that there exists an entire function h and entire functions f_1, g_1 such that $f = hf_1$, $g = hg_1$ and f_1, g_1 have no zeros in common.
(b) Show that there exist entire functions A, B such that $Af + Bg = h$.

Solution. (a) If f and g have no common zeros let $h = 1$. Otherwise let $\{z_1, \ldots\}$ be the set of common zeros of f and g. With the Weierstrass product we can construct and entire function h such that h has zeros at z_i for all i and such that the order of h at z_i is $\min(\mathrm{ord}_{z_i} f, \mathrm{ord}_{z_i} g)$. Then $f_1 = f/h$ and $g_1 = g/h$ are both entire functions with no common zeros.

(b) By Exercise 3 there exist entire functions A and B such that

$$Af_1 + Bg_1 = 1.$$

Hence

$$Af + Bg = h.$$

Exercise XIII.4.4. *Let* $f_1, \ldots, f_m$ *be a finite number of entire functions, and let* J *be the set of all combinations* $Af_1 + \cdots + A_m f_m$, *where* A_i *are entire functions. Show that there exists a single entire function* f *such that* J *consists of all multiples of* f, *that is,* J *consists of all entire functions* Af, *where* A *is entire. In the language of rings, this means that every finitely generated ideal in the ring of entire functions is principal.*

Solution. We prove the result by induction. By Exercise 3 there exists an entire function f such that $f_1 = f \tilde{f}_1$ and $f_2 = f \tilde{f}_2$ and $Af_1 + Bf_2 = f$ for some entire functions A and B. This implies that f is in J. Now given entire functions A_1 and A_2 we have

$$A_1 f_1 + A_2 f_2 = (A_1 \tilde{f}_1 + A_2 \tilde{f}_2)f = Af$$

for some entire function f. This proves the base step of the induction.

Suppose the result is true for $m - 1$ functions. Let g be a generator for the ideal generated by $f_1, \ldots, f_{m-1}$. By Exercise 3 there exists an entire function f such that $g = f \tilde{g}$ and $f_m = f \tilde{f}_m$ and $Ag + Bf_m = f$ for some entire functions A and B. So f belongs to the ideal generated by $f_1, \ldots, f_m$ and given entire functions $A_1, \ldots, A_m$ we get

$$A_1 f_1 + \cdots + A_m f_m = Cg + A_m f_m = (C\tilde{g} + A_m \tilde{f}_m)f = Af.$$

This proves that any finitely generated ideal of the ring of entire functions is principal. If f and g generate the same ideal, there exists entire functions A and B such that $Af = g$ and $Bg = f$. Therefore $AB = 1$.

Exercise XIII.4.5. *Let* $\{a_k\}, \{z_k\}$ *be sequences of nonzero complex numbers, with* $|z_k| \to \infty$ *and* $|z_k| \leq |z_{k+1}|$ *for all* k. *Let* ρ *be a real number* > 0 *such that*

$$\sum_{k=1}^{\infty} \frac{|a_k|}{|z_k|^\rho} < \infty.$$

Define

$$A_n = \sum_{k=1}^{n} |a_k|.$$

(a) Prove that $A_n = o(|z_n|^\rho)$ *for* $n \to \infty$, *meaning that* $\lim A_n/|z_n|^\rho = 0$.
(b) Let d *be the smallest integer* $\geq \rho$. *Let* G_d *be the polynomial*

$$G_d(z) = \sum_{n=0}^{d-1} z^n.$$

Define

$$F_k(z, z_k) = \frac{a_k}{z - z_k} + \frac{a_k}{z_k} G_d(z/z_k).$$

Prove that the series

$$F(z) = \sum_{k=1}^{\infty} F_k(z, z_k)$$

converges absolutely and uniformly on every compact set not containing any z_k.
(c) Let S be a subset of **C** *at finite nonzero distance from all z_k, that is, there exists $c > 0$ such that $|z - z_k| \geq c$ for all $z \in S$ and all k. Show that*

$$F(z) = O(|z|^d) \quad \text{for } z \in S, |z| \to \infty.$$

(d) Let U be the complement of the union of all discs $D(z_k, \delta_k)$, centered at z_k, of radius $\delta_k = 1/|z_k|^d$. Show that

$$F(z) = o(|z|^{\rho+d}) \quad \text{for } z \in U, |z| \to \infty.$$

Note: For part (d), you will probably need part (a), but for (c), you won't.

Solution. This result follows at once from Exercise 3, §1 in Chapter XVI. Indeed, let $c_k = |a_k|/|z_k|^\rho$ and $b_k = |z_k|^\rho$. Then

$$\lim_{n \to \infty} \frac{1}{b_n} \sum_{k=1}^{n} c_k b_k = 0$$

which is precisely what we want.
(b) Let K be a compact set not containing any z_k. Choose R so that K is contained in the ball of radius R centered at the origin and choose N so that $|z_N| > 2R$. Suppose $z \in K$ which implies $|z/z_k| < 1/2$ for $k \geq N$. Split the series in two parts

$$F(z) = \sum_{k=1}^{N} F_k(z, z_k) + \sum_{k=N+1}^{\infty} F_k(z, z_k).$$

It suffices to show that the second term converges absolutely. We may write

$$F_k(z, z_k) = \frac{-a_k}{z_k}\left(\frac{1}{1 - z/z_k}\right) + \frac{a_k}{z_k}\sum_{j=0}^{d-1}\left(\frac{z}{z_k}\right)^j$$

$$= \frac{-a_k}{z_k}\sum_{j=0}^{\infty}\left(\frac{z}{z_k}\right)^j + \frac{a_k}{z_k}\sum_{j=0}^{d-1}\left(\frac{z}{z_k}\right)^j.$$

Therefore

$$|F_k(z, z_k)| \leq \left|\frac{a_k}{z_k}\right|\sum_{j=d}^{\infty}\left|\frac{z}{z_k}\right|^j$$

$$\leq \left|\frac{a_k}{z_k}\right|\left|\frac{z}{z_k}\right|^d\sum_{j=0}^{\infty}\frac{1}{2^j}$$

$$\leq 2R^d \frac{|a_k|}{|z_k|^{d+1}}.$$

But $\sum |a_k|/|z_k|^{d+1} < \infty$, so we conclude that the series F converges absolutely and uniformly on compact sets not containing any z_k.

(c) We can write

$$
\begin{aligned}
F_k(z, z_k) &= \frac{a_k}{z - z_k} + \frac{a_k}{z_k} \left(\sum_{j=0}^{d-1} (z/z_k)^j \right) \\
&= \frac{a_k}{z - z_k} + \frac{a_k}{z_k} \left(\frac{1 - (z/z_k)^d}{1 - (z/z_k)} \right) \\
&= \frac{a_k(z/z_k)^d}{z - z_k}.
\end{aligned}
$$

Therefore

$$|F_k(z, z_k)| \leq \frac{|a_k| \, |z|^d}{c |z_k|^d},$$

and since $\sum |a_k|/|z_k|^d < \infty$ we conclude that $F(z) = O(|z|^d)$. Note that the standard method used in (b) to prove the convergence of $F(z)$ can be replaced by the estimates we just obtained, because any compact set not containing any z_k is at finite distance from $\{z_k\}_{k=1}^{\infty}$.

(d) Let $|z| = R$ be large, and choose N, depending on R so that $|z_N| \leq 2R$ and $|z_{N+1}| > 2R$. We estimate $|F_k(z, z_k)|$ by considering two cases, $k \leq N$ and $k > N$. For $k \leq N$, copying the computation in (c) and using the fact that $|z - z_k| \geq 1/|z_k|^d$ we find that

$$
\begin{aligned}
|F_k(z, z_k)| &\leq \frac{|a_k| \, |z|^d}{|z - z_k|} \\
&\leq \frac{|a_k| \, |z|^{\rho+d}}{|z|^{\rho}}.
\end{aligned}
$$

But $|z_N| \leq 2|z|$ so we conclude that for $k \leq N$ we have

$$|F_k(z, z_k)| \leq \frac{|a_k| \, |z|^{\rho+d}}{|z_N|^{\rho}},$$

and therefore we obtain

$$\sum_{k=1}^{N} |F_k(z, z_k)| \leq \frac{A_N}{|z_N|^{\rho}} |z|^{\rho+d}.$$

We now turn our attention to the estimate when $k > N$. In this case, copying part of the argument given in (b) we get

$$|F_k(z, z_k)| \leq \left| \frac{a_k}{z_k} \right| \left| \frac{z}{z_k} \right|^d \leq \frac{1}{|z|^{\rho}} \frac{|a_k|}{|z_k|^{d+1}} |z|^{\rho+d}.$$

Combining all these estimates, we see that if $|z| = R$ and if we denote the corresponding N by $N(R)$ we get

$$|F(z)| \leq \left(\frac{A_{N(R)}}{|z_{N(R)}|^\rho} + \frac{1}{|z|^\rho} \sum_{k=N(R)}^\infty \frac{|a_k|}{|z_k|^{d+1}} \right) |z|^{\rho+d}.$$

Since $N(R) \to \infty$ as $|z| \to \infty$, we see that we do have $F(z) = o(|z|^{\rho+d})$ as $|z| \to \infty, z \in U$.

XV
The Gamma and Zeta Functions

XV.1 The Differentiation Lemma

Exercise XV.1.1. *For* $\text{Re}(z) > 0$, *prove that*

$$\log z = \int_0^\infty (e^{-t} - e^{-zt})\frac{dt}{t}.$$

[Hint: Show that the derivatives of both sides are equal.]

Solution. If $\text{Re}(z) \geq \delta > 0$ and t is positive, then $|e^{-zt}| \leq e^{-t\delta}$. For t near zero we have

$$e^{-t} = 1 - t + \cdots$$

and

$$e^{-zt} = 1 - zt + \cdots$$

so we see that the integral converges uniformly for z is compact subsets of the right half plane. Differentiating we find that the derivative of the function defined by the integral is

$$\int_0^\infty e^{-zt}dt = \left[\frac{-1}{z}e^{-zt}\right]_0^\infty = \frac{1}{z}.$$

Evaluating $\int_0^\infty (e^{-t} - e^{-zt})\frac{dt}{t}$ at $z = 1$ we find 0, so we have the formula

$$\log z = \int_0^\infty (e^{-t} - e^{-zt})\frac{dt}{t}.$$

Exercise XV.1.2. *Let f be analytic on the closed unit disc. Let*

$$I(z) = \int_0^1 \frac{f(t)}{t+z} dt.$$

Show that $I(z) + f(-z) \log z$ is analytic for z in some neighborhood of 0. [Hint: First consider z real positive, or if you wish, z with positive real part. Use the power series expansion $f(t) = \sum c_k t^k$, and write $t = t + z - z$. Collect terms. The part

$$\sum_{k=0}^{\infty} c_k (-z)^k \int_0^1 \frac{dt}{t+z}$$

will give rise to the log term.]

Solution. Suppose that z has a positive real part. Then we write $f(t) = \sum c_k t^k$ so that the binomial expansion gives

$$f(t) = \sum c_k (t+z-z)^k = (t+z)h(z,t) + \sum c_k (-z)^k$$

where $h(z,t)$ is analytic for each t. Dividing by $t+z$ and integrating we see that the term on the right becomes

$$f(-z) \int_0^1 \frac{1}{t+z} dt = f(-z)[\log(1+z) - \log(z)].$$

For z near 0, $f(-z) \log(1+z)$ is analytic hence $I(z) + f(-z) \log(z)$ is analytic for z in some neighborhood of zero.

The Laplace Transform

Exercise XV.1.3. *Let f be a continuous function with compact support on the interval $[0, \infty[$. Show that the function Lf given by*

$$Lf(z) = \int_0^{\infty} f(t) e^{-zt} dt$$

is entire

Solution. Suppose that $|z| \le R$. Then we have $|e^{-zt}| \le e^{Rt}$, and since h has compact support, it is integrable and 0 outside some large interval. Since R was arbitrary we conclude that Lf is entire.

Exercise XV.1.4. *Let f be a continuous function on $[0, \infty[$, and assume that there is a constant $C > 1$ such that*

$$|f(t)| \ll C^t \quad for\ t \to \infty,$$

i.e., there exist constants A, B such that $|f(t)| \le Ae^{Bt}$ for all t sufficiently large. (a) Prove that the function

$$Lf(z) = \int_0^{\infty} f(t) e^{-zt} dt$$

is analytic in some half plane $\text{Re } z \geq \sigma$ *for some real number* σ. *In fact, the integral converges absolutely for some* σ. *Either such* σ *have no lower bound, in which case, Lf is entire, or the greatest lower bound* σ_0 *is called the **abscissa of convergence of the integral**, and the function Lf is analytic for* $\text{Re}(z) > \sigma_0$. *The integral converges absolutely for*

$$\text{Re } z \geq \sigma_0 + \epsilon,$$

for every $\epsilon > 0$.

*The function Lf is called the **Laplace transform** of f.*

(b) Assuming that f is of class C^1, *prove by integrating by parts that*

$$Lf'(z) = zLf(z) - f(0).$$

Solution. (a) Let $z = x + iy$. We estimate the integrand in the following way,

$$|f(t)e^{-zt}| \leq Ae^{Bt}e^{-tx}.$$

If $x \geq B + \epsilon$ where $\epsilon > 0$, the integrand is uniformly bounded by an integrable function, namely $Ae^{-\epsilon t}$, so the integral defines an analytic function on $\text{Re}(z) \geq B + \epsilon$ for every $\epsilon > 0$. Therefore Lf defines an analytic function on $\text{Re}(z) > B$.

(b) We integrate by parts $Lf(z)$ and get

$$Lf(z) = \left[\frac{-f(t)e^{-zt}}{z} \right]_0^\infty - \frac{1}{z} \int_0^\infty -f'(t)e^{-zt}dt$$

$$= \frac{f(0)}{z} + \frac{1}{z}Lf'(z).$$

Thus $Lf'(z) = zLf(z) - f(0)$ as was to be shown.

Find the Laplace transform of the following functions, and the abscissa of convergence of the integral defining the transform. In each case, a is a real number $\neq 0$.

Exercise XV.1.5. $f(t) = e^{-at}$.

Solution. We have

$$Lf(z) = \int_0^\infty e^{-at}e^{-zt}dt$$

so the abscissa of convergence of the integral is $\sigma_0 = -a$ and we have

$$Lf(z) = \left[\frac{-1}{a+z}e^{-(a+z)t} \right]_0^\infty = \frac{1}{a+z}.$$

Exercise XV.1.6. $f(t) = \cos at$.

Solution. The abscissa of convergence is 0 because $|\cos at| \leq 1$. We apply the formula obtained in part (b) of Exercise 4 to f' and we get

$$Lf''(z) = zLf'(z) - f'(0) = z^2Lf(z) - zf(0) - f'(0).$$

In this exercise we have $Lf''(z) = -a^2 Lf(z)$ so

$$-a^2 Lf(z) = z^2 Lf(z) - z,$$

and therefore

$$Lf(z) = \frac{z}{a^2 + z^2}.$$

Exercise XV.1.7. $f(t) = \sin at.$

Solution. Arguing like in Exercise 6, we find that the abscissa of convergence is 0 and that

$$Lf(z) = \frac{a}{a^2 + z^2}.$$

Exercise XV.1.8. $f(t) = (e^t + e^{-t})/2.$

Solution. From the inequality

$$|2f(t)e^{-zt}| = e^{t-tx} + e^{-t-tx}$$

we see that the abscissa of convergence is 1. Using the result of Exercise 5 we get

$$Lf(z) = \frac{1}{2} \left(\int_0^\infty e^t e^{-zt} dt + \int_0^\infty e^{-t} e^{-zt} dt \right) = \frac{1}{2} \left(\frac{1}{-1+z} + \frac{1}{1+z} \right)$$

$$= \frac{z}{z^2 - 1}.$$

Exercise XV.1.9. *Suppose that f is periodic with period $a > 0$, that is $f(t+a) = f(t)$ for all $t \geq 0$. Show that*

$$Lf(z) = \frac{\int_0^a e^{-zt} f(t)dt}{1 - e^{-az}} \quad \text{for } \operatorname{Re} z > 0.$$

Solution. We assume that f is integrable so that for $\operatorname{Re} z > \epsilon$ the integral is uniformly convergent and therefore defines an analytic function on the right half plane. We can write the integral as an infinite sum and use the periodicity of f to get

$$Lf(z) = \int_0^a f(t)e^{-zt} dt + \int_a^{2a} f(t)e^{-zt} dt + \cdots + \int_{na}^{(n+1)a} f(t)e^{-zt} dt + \cdots$$

$$= \int_0^a f(u)e^{-zu} du + \int_0^a f(u)e^{-z(u+a)} du + \cdots$$

$$+ \int_0^a f(u)e^{-z(u+na)} du + \cdots$$

$$= \int_0^a f(u)e^{-zu}(1 + e^{-za} + \cdots)du.$$

But

$$1 + e^{-za} + \cdots + e^{-(N-1)za} = \frac{1 - e^{-Nza}}{1 - e^{-za}},$$

and since $|e^{-Nza}| = e^{-N\,\text{Re}(z)a} \to 0$ as $N \to \infty$ we get

$$Lf(z) = \frac{\int_0^\infty f(u)e^{-zu}du}{1 - e^{-za}}$$

for $\text{Re}(z) > 0$, as was to be shown.

XV.2   The Gamma Function

Exercise XV.2.1. *Prove that:*
(a) $\Gamma'/\Gamma(1) = -\gamma$.
(b) $\Gamma'/\Gamma(\frac{1}{2}) = -\gamma - 2\log 2$.
(c) $\Gamma'/\Gamma(2) = -\gamma + 1$.

Solution. (a) The result follows from the formula $\Gamma 2$ namely

$$-\Gamma'/\Gamma(1) = 1 + \gamma + \sum_{n=1}^\infty \left(\frac{1}{1+n} - \frac{1}{n}\right) = 1 + \gamma - 1.$$

(b) By $\Gamma 2$ we get

$$\Gamma'/\Gamma\left(\frac{1}{2}\right) = 2+\gamma+2\sum_{n=1}^\infty\left(\frac{1}{1+2n} - \frac{1}{2n}\right) = \gamma + 2\sum_{n=1}^\infty \frac{(-1)^{n+1}}{n} = \gamma + 2\log 2.$$

(c) By $\Gamma 2$ we get

$$-\Gamma'/\Gamma(2) = \frac{1}{2} + \gamma + \sum_{n=1}^\infty\left(\frac{1}{2+n} - \frac{1}{n}\right) = \frac{1}{2} + \gamma - 1 - \frac{1}{2}.$$

Exercise XV.2.2. *Give the details for the proofs of formulas $\Gamma 10$ and $\Gamma 11$*

Solution. In Exercise 1 (a) we showed that $\Gamma'/\Gamma(1) = -\gamma$, so putting $z = 1$ in $\Gamma 9$ we get

$$\int_0^\infty \left(\frac{e^{-t}}{t} - \frac{e^{-t}}{1 - e^{-t}}\right) dt = -\gamma,$$

whence

$$\int_0^\infty \left(\frac{1}{1 - e^{-t}} - \frac{1}{t}\right) e^{-t}dt = \gamma.$$

Using this formula we see that

$$-\gamma + \int_0^\infty \frac{e^{-t} - e^{-tz}}{1 - e^{-t}}dt = \int_0^\infty \left(\frac{e^{-t}}{t} - \frac{-e^{-tz}}{1 - e^{-t}}\right) dt = \Gamma'/\Gamma(z).$$

Exercise XV.2.3. *Prove that $\int_0^\infty e^{-t}\log t\,dt = -\gamma$.*

Solution. For $\text{Re}(z) > 0$ we have $\Gamma(z) = \int_0^\infty e^{-t}t^z \frac{dt}{t}$, thus

$$\Gamma'(z) = \int_0^\infty e^{-t}(\log t)t^z \frac{dt}{t}.$$

Therefore

$$\Gamma'(1) = \int_0^\infty e^{-t}(\log t)dt,$$

but $\Gamma(1) = 1$, so from Exercise 1 we get $\Gamma'(1) = -\gamma$ so

$$\int_0^\infty e^{-t}(\log t)dt = -\gamma$$

as was to be shown.

Exercise XV.2.4. *Show that*

$$\int_0^1 \left(\frac{1}{e^t - 1} - \frac{1}{t}\right)dt + \int_1^\infty \frac{dt}{e^t - 1} = 0.$$

Solution. From $\Gamma 9$ and Exercise 1 (a) we have

$$\int_0^\infty \left(\frac{e^{-t}}{t} - \frac{1}{e^t - 1}\right)dt = -\gamma.$$

Split the integral from 0 to 1 and from 1 to ∞. Integration by parts and Exercise 3 imply

$$\int_1^\infty \frac{e^{-t}}{t}dt = \int_1^\infty e^{-t}\log t\,dt = -\gamma - \int_0^1 e^{-t}\log t\,dt.$$

So we have shown that

$$\int_0^1 \left(\frac{e^{-t}}{t} - e^{-t}\log t - \frac{1}{e^t - 1}\right)dt - \int_1^\infty \frac{dt}{e^t - 1} = 0.$$

We now investigate the first integral on the left. Let $0 < \delta < 1$. Then integrating by parts we find that

$$\int_\delta^1 e^{-t}\log t\,dt = e^{-\delta}\log\delta + \int_\delta^1 \frac{e^{-t}}{t}dt = -e^{-\delta}\int_\delta^1 \frac{dt}{t} + \int_\delta^1 \frac{e^{-t}}{t}dt.$$

so

$$\int_\delta^1 \left(\frac{e^{-t}}{t} - e^{-t}\log t - \frac{1}{e^t - 1}\right)dt = \int_\delta^1 \left(e^{-\delta}\frac{1}{t} - \frac{1}{e^t - 1}\right)dt$$

$$= \int_\delta^1 \left(\frac{1}{t} - \frac{1}{e^t - 1}\right)dt + \int_\delta^1 \frac{e^{-\delta} - 1}{t}dt.$$

Now let $\delta \to 0$. The last integral tends to 0 because

$$\int_\delta^1 \frac{e^{-\delta} - 1}{t}dt = (e^{-\delta} - 1)\log\delta.$$

The desired formula now drops out.

Exercise XV.2.5. *Let* $a_1, \ldots, a_r$ *be distinct complex numbers, and let* $m_1, \ldots, m_r$ *be integers. Suppose that*

$$h(z) = \prod_{i=1}^{r} \Gamma(z + a_i)^{m_i}$$

is an entire function without zeros and poles.
(a) Prove that there are constants A, B *such that* $h(z) = AB^z$.
(b) Assuming (a), prove that that $m_i = 0$ *for all* i.

Solution. (a) We may write

$$h(z) = \prod_{i=1}^{r} \Gamma(z + a_i)^{m_i} = \frac{\prod_{m_i \geq 0} \Gamma(z + a_i)^{m_i}}{\prod_{m_i < 0} \Gamma(z + a_i)^{|m_i|}}.$$

We know that $1/\Gamma(z)$ has order 1, so the two functions

$$\frac{1}{\prod_{m_i \geq 0} \Gamma(z + a_i)^{m_i}} \quad \text{and} \quad \frac{1}{\prod_{m_i < 0} \Gamma(z + a_i)^{|m_i|}}$$

also have order 1. By Exercise 2, §3 Chapter XIII, we find that h has order 1. By Hadamard's theorem, we conclude that $h(z) = e^{az+b}$ for some $a, b \in \mathbf{C}$.
(b) Let $h(z) = \prod_{i=1}^{r} \Gamma(z + a_i)^{m_i}$. Then the logarithmic derivative of h is

$$h'/h(z) = \sum_{i=1}^{r} m_i \Gamma' / \Gamma(z + a_i)$$

and we know that

$$-\Gamma'/\Gamma(z) = \frac{1}{z} + \gamma + \sum_{n=1}^{\infty} \left(\frac{1}{z+n} - \frac{1}{n} \right)$$

so the set of poles of $\Gamma'/\Gamma(z + a_i)$ is

$$P_i = \{-a_i - n : n = 0, 1, 2, \ldots\}$$

and the residue at the poles P_i is $-m_i$. Let $P = \bigcup_{i=1}^{r} P_i$. Since we assume that $h(z) = e^{A+Bz}$ we must have $h'/h(z) = B$. Hence all the poles cancel. We must show that this implies $m_i = 0$ for all i. After renumbering the $a_j's$ we may assume that Re $-a_{j+1} \leq$ Re $-a_j$ for all j. Let $b_j = -a_j$. We can find a small circle C_1 centered at b_1 containing no other point of P. Cauchy's theorem implies

$$\int_{C_1} h'/h = 2\pi i(-m_1).$$

But $h'/h = B$ so the integral is 0 and therefore $m_1 = 0$. Now we proceed by induction. Suppose that $m_1 = \cdots = m_k = 0$. Consider a small circle C_{k+1} around b_{k+1}. When we integrate h'/h over C_{k+1} we must consider two cases. If $b_{k+1} \notin P_i$ for all $1 \leq i \leq k$, the residue is $-m_{k+1}$ so we find $m_{k+1} = 0$. If $b_{k+1} \in P_i$ for some $0 \leq i \leq k$ then the residue is

$$-m_{k+1} - \sum m_i.$$

where the sum is taken over some i's with $1 \leq i \leq k$. The induction hypothesis implies that the sum is 0 and therefore $m_{k+1} = 0$ as was to be shown.

Exercise XV.2.6. *(a) Give an exact value for $\Gamma(1/2 - n)$ when n is a positive integer, and thus show that $\Gamma(1/2 - n) \to 0$ rapidly when $n \to \infty$. Thus the behavior at half the odd integers is quite opposite to the polar behavior at the negative integers themselves.*
(b) Show that $\Gamma(1/2 - n + it) \to 0$ uniformly for real t, as $n \to \infty$, n equal to a positive integer.

Solution. To give an exact value of $\Gamma(1/2 - n)$ we use the fact that $\Gamma(z + 1) = z\Gamma(z)$. By induction, we prove that for $n \geq 1$ we have

$$\Gamma(1/2 - n) = \frac{(-2)^n}{(2n - 1) \cdots 5 \times 3 \times 1}\Gamma(1/2) = \frac{(-2)^n 2^n (n!)}{(2n)!}\Gamma(1/2).$$

The formula is true for $n = 1$ because

$$(-1/2)\Gamma(-1/2) = \Gamma(1/2)$$

which implies

$$\Gamma(-1/2) = -2\Gamma(1/2) = \frac{(-2) \times 2}{2!}\Gamma(1/2).$$

Also, we have

$$(1/2 - (n + 1))\Gamma(1/2 - (n + 1)) = \Gamma(1/2 - n).$$

So

$$\Gamma(1/2 - (n + 1)) = \frac{-2}{2(n + 1) - 1}\frac{(-2)^n}{(2n - 1) \cdots 5 \times 3 \times 1}\Gamma(1/2)$$

$$= \frac{(-2)^{n+1}}{(2(n + 1) - 1) \cdots 5 \times 3 \times 1}\Gamma(1/2)$$

$$= \frac{(-2)^{n+1} 2^{n+1}(n + 1)!}{(2(n + 1))!}\Gamma(1/2)$$

which concludes the induction. Of course, we may replace $\Gamma(1/2)$ by its value $\sqrt{\pi}$. So we have

$$|\Gamma(1/2 - n)| \leq C\frac{4^n n!}{(2n)!}.$$

Let $a_n = 4^n n!/(2n)!$. It suffices to show that a_n decreases rapidly to 0 as $n \to \infty$. Indeed,

$$\frac{a_{n+1}}{a_n} = \frac{4(n + 1)}{(2n + 2)(2n + 1)} = \frac{2}{(2n + 1)} \leq \frac{2}{3} < 1$$

whenever $n \geq 1$, so that $a_n = O((2/3)^n)$ as $n \to \infty$.
(b) Arguing like above and noting that $|1 - 2n + 2it| \geq |1 - 2n|$ we find

$$|\Gamma(1/2 - n + it)| \leq \frac{4^n (n!)}{(2n)!}|\Gamma(1/2 + it)|.$$

It is therefore sufficient to show that $\Gamma(1/2 + it)$ is uniformly bounded for all real t. This follows from Stirling's formula. Indeed,

$$\Gamma(1/2 + it) \sim (1/2 + it)^{it} e^{-1/2 - it} \sqrt{2\pi},$$

and

$$|(1/2 + it)^{it} e^{-1/2 - it}| \leq |e^{it \log(1/2 + it)}| = e^{-t\theta_t}$$

where θ_t denotes the argument of $1/2 + it$. Since t and θ_t are of the same sign, we conclude that $\Gamma(1/2 + it)$ is uniformly bounded for real t. This concludes the exercise.

Exercise XV.2.7. *Mellin Inversion Formula.* *Show that for $x > 0$ we have*

$$e^{-x} = \frac{1}{2\pi i} \int_{\sigma = \sigma_0} x^{-s} \Gamma(s)\, ds,$$

where $s = \sigma + it$, and the integral is taken over a vertical line with fixed real part $\sigma_0 > 0$ and $-\infty < t < \infty$. [Hint: What is the residue of $x^{-s}\Gamma(s)$ at $s = -n$?]

Solution. The residue of $x^{-s}\Gamma(s)$ at $s = -n$ is

$$x^n \frac{(-1)^n}{n!} = \frac{(-x)^n}{n!}.$$

We now prove the Mellin inversion formula, using the calculus of residues. Let $\alpha_n = (1/2) - n$. Consider the rectangle with corners

$$\sigma_0 + iT, \quad \alpha_n + iT, \quad \alpha_n - iT \quad \text{and} \quad \sigma_0 - iT.$$

Denote by $R_{n,T}$ this rectangle. Then,

$$\frac{1}{2\pi i} \int_{R_{n,T}} x^{-s} \Gamma(s)\, ds = \sum_{j=0}^{n-1} \frac{(-x)^j}{j!}.$$

We first show that the contribution of this integral over the horizontal segments goes to 0 as $T \to \infty$. Write $s = \sigma + it$ with $|\sigma|$ bounded and $t = T$. By Stirling's formula we have

$$\Gamma(s) \sim s^{s-1/2} e^{-s} \sqrt{2\pi} \quad \text{as } |s| \to \infty.$$

But

$$s^{s-1/2} e^{-s} = e^{(\sigma + it - 1/2)(\log|s| + i\arg(s))} e^{-\sigma - it}$$

hence

$$|s^{s-1/2} e^{-s}| = |s|^{\sigma - 1/2} e^{-t\arg(s)} e^{-\sigma}.$$

However, t and $\arg(s)$ have the same sign, so if s is contained in a vertical strip and $|t| \geq 1$ we have

$$\Gamma(s) = O\left(|t|^{\sigma - 1/2} e^{-|t|}\right).$$

We conclude from this estimate that the integral of $\Gamma(s)x^{-s}$ over the horizontal segments of the rectangle goes to 0 as $T \to \infty$ with n fixed. To conclude the proof, it suffices to show that the integral of $\Gamma(s)x^{-s}$ over the vertical line $\text{Re}(s) = -n + 1/2$ goes to 0. To show this, we can use Exercise 6 (b). First, a change of variable gives

$$\int_{\text{Re}(s)=-n+1/2} \Gamma(s)x^{-s}ds = \int_{\text{Re}(\zeta)=1/2} \Gamma(\zeta-n)x^{-\zeta+n}d\zeta.$$

But we know from Exercise 6 (b) that

$$|\Gamma(1/2-n+it)| \leq \frac{4^n(n!)}{(2n)!}|\Gamma(1/2+it)|,$$

Therefore

$$\left|\int_{\text{Re}(s)=-n+1/2} \Gamma(s)x^{-s}ds\right| \leq \frac{4^n(n!)}{(2n)!}x^{-1/2+n}\int_{-\infty}^{\infty}|\Gamma(1/2+it)|dt.$$

Stirling's formula (see Exercise 6 (b)) shows that Γ is integrable over the line $\text{Re}(s) = 1/2$, and this proves that the desired integral goes to 0. Hence

$$e^{-x} = \frac{1}{2\pi i}\int_{\sigma=\sigma_0} x^{-s}\Gamma(s)ds$$

as was to be shown.

Exercise XV.2.8. *Define the alternate Laplace transform L^- by*

$$L^- f(w) = \int_0^\infty f(t)e^{wt}dt.$$

(a) Let $f(t) = e^{-zt}$ for $t \geq 0$. Show that

$$L^- f(w) = \frac{1}{z-w} \quad \text{for } \text{Re}(w) < \text{Re}(z).$$

(b) Let $f(t) = t^{s-1}e^{-zt}$ for $t \geq 0$. Show that

$$L^- f(w) = \Gamma(s)(z-w)^{-s} \quad \text{for } \text{Re}(w) < \text{Re}(z).$$

Here $(z-w)^s$ is defined by taking $-\pi/2 < \arg(z-w) < \pi/2$.

Solution. (a) For $\text{Re}(w) < \text{Re}(z)$ we have

$$\int_0^\infty e^{-zt}e^{wt}dt = \frac{1}{-(z-w)}\left[e^{-(z-w)t}\right]_0^\infty = \frac{1}{z-w}.$$

(b) Consider the function $g(\zeta) = \zeta^{s-1}e^{-\zeta}$ for the logarithm defined on $\mathbf{C} - \mathbf{R}_{\leq 0}$. integrating g along the contour given by

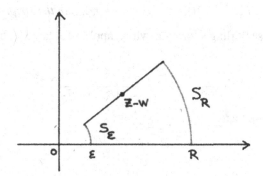

we obtain 0 by Cauchy's formula. We claim that the integral along the arcs tend to 0. For the large arc S_R we have

$$I(R) = \int_{S_R} g(\zeta)d\zeta = \int_0^\varphi (Re^{i\theta})^{s-1}e^{-Re^{i\theta}} Rie^{i\theta}d\theta.$$

Here, $S_R = \{Re^{i\theta} : 0 \le \theta \le \varphi\}$. Choose $c > 0$ so that for $0 \le \theta \le \varphi$ we have $c \le \cos\theta \le 1$. Then

$$|I(R)| \le \int_0^\varphi R^{\mathrm{Re}(s)-1}e^{-\theta\,\mathrm{Im}(s)}e^{-R\cos\theta} Rd\theta$$

$$\le C R^{\mathrm{Re}(s)}e^{-Rc}$$

so that $|I(R)| \to 0$ as $R \to \infty$. Similarly, the integral along the arc of radius ϵ is bounded by $C'\epsilon^{\mathrm{Re}(s)}e^{-\epsilon c}$ which goes to 0 as $\epsilon \to 0$. Therefore, letting $R \to \infty$ and $\epsilon \to 0$ we get

$$\int_0^\infty g(\zeta)d\zeta = \int_L g(\zeta)d\zeta$$

where L is the line segment $u(z - w)$ with $0 \le u < \infty$ and oriented by increasing u. We obtain

$$\int_0^\infty t^{s-1}e^{-t}dt = \int_0^\infty ((z - w)u)^{s-1}e^{-(z-w)u}(z - w)du$$

hence

$$\Gamma(s) = (z - w)^s L^- f(w)$$

as was to be shown.

Exercise XV.2.9. *Consider the gamma function in a vertical strip* $x_1 \le \mathrm{Re}(z) \le x_2$. *Let a be a complex number. Show that the function*

$$z \mapsto \Gamma(z + a)/\Gamma(z) = h(z)$$

has a polynomial growth in the strip (as distinguished from exponential growth). In other words, there exists $k > 0$ such that

$$|h(z)| = O(|z|^k) \quad \text{for } |z| \to \infty, \text{ } z \text{ in the strip.}$$

Solution. We use Stirling's formula, which applies for large $|z|$ because we are in a strip. We have

$$\frac{\Gamma(z+a)}{\Gamma(z)} \sim \frac{(z+a)^{z+a-1/2}e^{-z+a}\sqrt{2\pi}}{z^{z-1/2}e^{-z}\sqrt{2\pi}}.$$

It suffices to prove that

$$\frac{(z+a)^{z+a-1/2}}{z^{z-1/2}}$$

has polynomial growth. Write $z = x+iy$ and $a = u+iv$ and let θ be the argument of $z + a$. Also, let φ be the argument of z. Then

$$|(z+a)^{z+a-1/2}| = |e^{(z+a-1/2)\log(z+a)}| = e^{(x+u-1/2)\log|z+a|-\theta(y+v)}$$

and

$$|z^{z-1/2}| = |e^{(z-1/2)\log z}| = e^{(x-1/2)\log|z|-y\varphi}.$$

As $|z| \to \infty$, we have $|z| \sim |z+a|$ and since we assume that x is bounded we have

$$|(z+a)^{z+a-1/2}| \sim C_1 e^{C_2 \log|z|+D|z|} \quad \text{and} \quad |z^{z-1/2}| \sim B_1 e^{B_2 \log|z|+D|z|}$$

where C_1, C_2, B_1, B_2 and D are constants. This proves that the quotient $\Gamma(z+a)/\Gamma(z)$ has polynomial growth in a strip.

   Let the **Paley–Wiener space** consist of those entire functions f for which there exists a positive number C having the following property. Given an integer $N > 0$, we have

$$|f(x+iy)| \ll \frac{C^{|x|}}{(1+|y|)^N},$$

where the implied constant in $\ll$ depends on f and N. We may say that f is at most of exponential growth with respect to x, and is rapidly decreasing, uniformly in every vertical strip of finite width.

Exercise XV.2.10. If f is C^∞ (infinitely differentiable) on the open interval $]0, \infty[$ and has compact support, then its **Mellin transform** Mf defined by

$$Mf(z) = \int_0^\infty f(t)t^z \frac{dt}{t}$$

is in the Paley–Wiener space. [Hint: Integrate by parts.]

Solution. Select $0 < a < 1 < b < \infty$ such that the support of f is contained in $[a, b]$. Since f has compact support we can apply Lemma 1.1 and we see that Mf

is entire. Integrating by parts $n + 1$ times we find

$$Mf(z) = \frac{(-1)^{n+1}}{z(z+1)\cdots(z+n)} \int_0^\infty f^{n+1}(t)t^{z+n} dt.$$

We have the estimate

$$\left| \int_0^\infty f^{n+1}(t)t^{z+n} dt \right| \le \int_a^b |f^{n+1}(t)| \, |t^{z+n}| dt \le b^{|x|}b^n \int_a^b |f^{n+1}(t)| dt.$$

Conclude.

Exercise XV.2.11. *Let F be in the Paley–Wiener space. For any real x, define the function $\,'M_x F$ by*

$$'M_x F(t) = \int_{\mathrm{Re}(z)=x} F(z)t^z \frac{dt}{i}.$$

The integral is supposed to be taken on the vertical line $z = x + iy$, with fixed x, and $-\infty < y < \infty$. Show that $\,'M_x F$ is independent of x, so can be written $\,'MF$. [Hint: Use Cauchy's theorem.] Prove that $\,'M_x F$ has compact support on $]0, \infty[$.

Solution. Let x_0 and x_1 be real numbers. We integrate $F(z)t^z$ over a rectangle as shown on the figure:

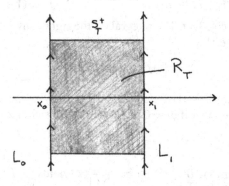

Since F is entire, Cauchy's theorem implies

$$\int_{R_T} F(z)t^z \frac{dt}{i} = 0.$$

We are interested in the behavior of the integral over the horizontal segments as $T \to \infty$. Consider the segment in the upper half plane, call it S_T^+ and parametrize it by $u + iT$ with $x_0 \le u \le x_1$. We then have

$$\left| \int_{S_T^+} F(z)t^z \frac{dt}{i} \right| \le \int_{x_0}^{x_1} |F(u + iT)| \, t^u du.$$

But F belongs to the Paley–Wiener space so

$$\left| \int_{S_T^+} F(z)t^z \frac{dt}{i} \right| \leq M \int_{x_0}^{x_1} \frac{C^{|u|}}{(1+|T|)} t^u \, du \leq \frac{M'}{(1+|T|)},$$

where M and M' are positive constants. This proves that the integral along S_T^+ tends to 0 as $T \to \infty$. The same result holds for the segment in the lower half plane, so combined with Cauchy's theorem and the correct orientation we find

$$\int_{\text{Re}(z)=x_0} F(z)t^z \frac{dz}{i} - \int_{\text{Re}(z)=x_1} F(z)t^z \frac{dz}{i} = 0.$$

Hence ${}^t M_{x_0} F(t) = {}^t M_{x_1} F(t)$, as was to be shown.

We now prove that ${}^t M_x F$, has compact support in $]0, \infty[$. Choose $N = 2$ and let C be the constant that appears in the Paley–Wiener estimate. We may assume of course that $C > 1$. Suppose that $0 < t \leq 1/(2C)$. Then the fact that

$$\int_{\text{Re}(\sigma)} F(z)t^z \frac{dz}{i} = \int_{-\infty}^{\infty} F(\sigma + iu)t^{\sigma+iu} \, du$$

implies the following estimate

$$|{}^t M F| \leq B \int_{\infty}^{\infty} \frac{C^{|\sigma|}}{(1+|u|)^2} t^\sigma \, du \leq BC^{|\sigma|} t^\sigma \int_{-\infty}^{\infty} \frac{1}{(1+|u|)^2} du,$$

where B is a positive constant. The integral on the right converges, so there exists a constant B' which verifies

$$|{}^t M F| \leq B' \frac{1}{2^\sigma}$$

for all $\sigma > 0$. Letting $\sigma \to \infty$ yields ${}^t M F = 0$. Now suppose $t \geq 2C$. Then a similar argument shows that if $\sigma < 0$, then we get the estimate

$$|{}^t M F| \leq B'' \frac{1}{2^{|\sigma|}}.$$

for some positive constant B'. Letting $\sigma \to -\infty$ yields ${}^t M F = 0$. Hence ${}^t M F$ has compact support in $]0, \infty[$.

Exercise XV.2.12. *Let a, b be real numbers > 0. Define the K-Bessel function*

K 1. $$K_s(a, b) = \int_0^\infty e^{-(a^2 t + b^2/t)} t^s \frac{dt}{t}.$$

Prove that K_s is an entire function of s. Prove that

K 2. $$K_s(a, b) = (b/a)^s K_s(ab),$$

where for $c > 0$ we define

K 3. $$K_s(c) = \int_0^\infty e^{-c(t+1/t)} t^s \frac{dt}{t}.$$

Prove that

K 4. $K_s(c) = K_{-s}(c).$

K 5. $K_{1/2}(c) = \sqrt{\pi/c}\, e^{-2c}.$

[Hint: Differentiate the integral for $\sqrt{x}\,K_{1/2}(x)$ under the integral sign.] Let $x_0 > 0$ and $\sigma_0 \leq \sigma \leq \sigma_1$. Show that there exists a number $C = C(x_0, \sigma_0, \sigma_1)$ such that if $x \geq x_0$, then

K 6. $K_\sigma \leq Ce^{-2x}.$

Prove that

K 7. $\displaystyle\int_{-\infty}^{\infty} \frac{1}{(u^2+1)^s}\,du = \sqrt{\pi}\,\frac{\Gamma(s-1/2)}{\Gamma(s)}$

for $\mathrm{Re}(s) > 1/2$. *Also prove that*

K 8. $\displaystyle\Gamma(s)\int_{-\infty}^{\infty} \frac{e^{ixu}}{(u^2+1)^s}\,du = 2\sqrt{\pi}\,(x/2)^{s-1/2}K_{s-1/2}(x)$

for $\mathrm{Re}(s) > 1/2$.

Solution. If $s = x + iy$ belongs to a compact set K, we want to show that the two integrals

$$\int_0^1 e^{-(a^2t+b^2/t)}t^{x-1}\,dt \quad \text{and} \quad \int_1^\infty e^{-(a^2t+b^2/t)}t^{x-1}\,dt$$

converge uniformly in s. The first integral converges because near 0, e^{-a^2t} is bounded and $\int_0^1 e^{-b^2/t}t^{x-1}\,dt$ converges uniformly for $s \in K$ because for all small t we have $e^{-b^2/t}t^{x-1} \leq e^{-b^2/2t}$.

The second integral also converges because for all large t, $e^{-b^2/t}$ is bounded and for all large t and all $s \in K$ we have $e^{-a^2t}t^{x-1} \leq e^{-a^2t/2}$. Since the integrand $(e^{-(a^2t+b^2/t)}t^s)/t$ is homomorphic in s for each $t > 0$ we conclude that $K_s(a, b)$ is entire.

K 3. In the integral $K_s(a, b)$ put $t = (bu)/a$. Then

$$K_s(a, b) = \int_0^\infty e^{-(abu+ab/u)}u^s \left(\frac{b}{a}\right)^s \frac{du}{u} = \left(\frac{b}{a}\right)^s K_s(ab).$$

K 4. We change variables $u = 1/t$, then

$$K_{-s}(c) = \int_\infty^0 e^{-c(1/u+u)}u^s u(-u^{-2})\,du = \int_0^\infty e^{-c(u+1/u)}u^s \frac{du}{u} = K_s(c).$$

K 5. Let $g(x) = K_{1/2}(x)$ and $h(x) = \sqrt{x}\,g(x)$. Changing variables $t = u/x$ we get

$$g(x) = \int_0^\infty e^{-u-x^2/u}u^{-1/2}x^{-1/2}\,du$$

so

$$h(x) = \int_0^\infty e^{-u-x^2/u}u^{-1/2}\,du.$$

Since

$$\frac{\partial}{\partial x}(e^{-u-x^2/u}u^{-1/2}) = \frac{-2x}{u^{3/2}}e^{-u-x^2/u},$$

we see that the integral

$$\int_0^\infty D_2(e^{-u-x^2/u}u^{-1/2})du$$

converges uniformly for $x \in [a, b]$ with $0 < a < b$, so we can differentiate under the integral sign and we get

$$h'(x) = \int_0^\infty \frac{-2x}{u^{3/2}}e^{-u-x^2/u}du.$$

The change of variables $q = x^2/u$ gives

$$h'(x) = \int_\infty^0 \frac{-2x}{(x^2/q)^{3/2}}e^{-q-x^2/q}(-x^2/q^2)dq = -2h(x),$$

so $h(x) = Ce^{-2x}$ for some constant C. Moreover, if we let $u = \alpha^2$ and if we use the continuity of h at 0 we get

$$C = \int_0^\infty e^{-u}u^{-1/2}du = 2\int_0^\infty e^{-\alpha^2}\alpha^{-1}\alpha d\alpha = \sqrt{\pi},$$

so

$$K_{1/2}(c) = \sqrt{\frac{\pi}{c}}e^{-2c}.$$

K 6. By definition we have

$$K_\sigma(x) = \int_0^\infty e^{-x(t+1/t)}t^s\frac{dt}{t}.$$

We split this integral as

$$\int_0^\infty = \int_0^{1/8} + \int_{1/8}^8 + \int_8^\infty$$

and get the desired bound for each integral separately. Since $t + 1/t \geq 2$ for all $t > 0$ we see that the middle integral is trivially bounded by a constant times e^{-2x}. For the first integral, note that

$$\frac{1}{t} \geq 2 + \frac{1}{2t}$$

whenever $0 < t \leq 1/8$. Therefore

$$\int_0^{1/8} e^{-x(t+1/t)}t^s\frac{dt}{t} \leq e^{-2x}\int_0^{1/8} e^{-x(t+1/2t)}t^s\frac{dt}{t} \leq Ce^{-2x}.$$

A similar argument applies to the third integral.

K 7. We have

$$\Gamma(s) \int_{-\infty}^{\infty} \frac{1}{(u^2+1)^s} du = 2 \int_0^{\infty} \int_0^{\infty} \frac{e^{-t} t^{s-1}}{(u^2+1)^s} dt\, du.$$

If $t = (u^2+1)q$, then $dt = (u^2+1)dq$ so that

$$\Gamma(s) \int_{-\infty}^{\infty} \frac{1}{(u^2+1)^s} du = 2 \int_0^{\infty} \int_0^{\infty} e^{-u^2 q} e^{-q} q^{s-1} dq\, du.$$

The change of variable $\alpha = u\sqrt{q}$ implies

$$\int_0^{\infty} e^{-u^2 q} du = \frac{1}{\sqrt{q}} \int_0^{\infty} e^{-\alpha^2} d\alpha = \frac{\sqrt{\pi}}{2} q^{-1/2}.$$

For $\mathrm{Re}(s) > 1/2$ the hypotheses of Theorem 3.5 are verified so that

$$\Gamma(s) \int_{-\infty}^{\infty} \frac{1}{(u^2+1)^s} du = 2 \int_0^{\infty} \int_0^{\infty} \frac{e^{-u^2 q} e^{-q} q^{s-1}}{(u^2+1)^s} du\, dq$$

$$= \sqrt{\pi} \int_0^{\infty} e^{-q} q^{s-1-1/2} \frac{dq}{q} = \sqrt{\pi}\, \Gamma\left(s - \frac{1}{2}\right).$$

K 8. To prove this last formula, proceed like we just did. Write $\Gamma(s)$ as an integral, change the order of integration, and make the change of variables $t \to (u^2 + 1)t$. Then use the fact that $e^{-x^2/2}$ is its own Fourier transform where the Fourier transform is appropriately normalized. Make a final change of variables $t \to tx$ and conclude.

XV.3   The Lerch Formula

Exercise XV.3.1. *For each real number x, we let $\{x\}$ be the unique number such that $x - \{x\}$ is an integer and $0 < \{x\} \le 1$. Let N be a positive integer. Prove the addition formula (distribution relation)*

$$N^{-s} \sum_{j=0}^{N-1} \zeta(s, \{x + j/N\}) = \zeta(s, \{Nx\}).$$

From this formula and Theorem 3.2, deduce another proof for the multiplication formula of the gamma function.

Solution. Fix x and select j' such that $\{x + j/N\} = \{x\} + j/N$ for all $j \le j'$ and $\{x + j/N\} = \{x\} + j/N - 1$ for all $j > j'$. Also, let ρ be the unique integer such that

$$\frac{\rho}{N} \le \{x\} < \frac{\rho+1}{N}.$$

Then, $N\{x\} = \{Nx\} + \rho$.

Now write the sum $N^{-s} \sum_{j=0}^{N-1} \zeta(s, \{x + j/N\})$ as a double sum

$$N^{-s} \sum_{j=0}^{N-1} \zeta(s, \{x + j/N\}) = N^{-s} \sum_{j=0}^{N-1} \sum_{n=0}^{\infty} \frac{1}{(n + \{x + j/N\})^s}$$

and interchange sum signs to get

$$\sum_{n=0}^{\infty} \sum_{j=0}^{N-1} \frac{1}{N^s(n + \{x + j/N\})^s}.$$

We now split the inner sum in two parts, namely

$$\sum_{j=0}^{j'} \frac{1}{N^s(n + \{x + j/N\})^s} + \sum_{j=j'+1}^{N-1} \frac{1}{N^s(n + \{x + j/N\})^s}.$$

Some simple manipulations show that the first sum is equal to

$$\sum_{j=0}^{j'} \frac{1}{(nN + \{Nx\} + \rho + j)^s}$$

and the second sum is equal to

$$\sum_{j=j'+1}^{N-1} \frac{1}{((n-1)N + \{Nx\} + \rho + j)^s}.$$

Now collecting terms properly, we find that

$$N^{-s} \sum_{j=0}^{N-1} \zeta(s, \{x + j/N\}) - \zeta(s, \{Nx\})$$

is equal to

$$\sum_{j=j'+1}^{N-1} \frac{1}{(-N + \{Nx\} + \rho + j)^s} - \frac{1}{(\{Nx\})^s} - \frac{1}{(1 + \{Nx\})^s} - \cdots - \frac{1}{(\rho - 1 + \{Nx\})^s}.$$

To show that the above expression is equal to 0, it suffices to notice that $\rho + j' + 1 = N$. This proves the addition formula.

To prove the multiplication formula of the gamma function,

$$\prod_{j=0}^{N-1} D\left(z + \frac{j}{N}\right) = D(Nz)N^{Nz-1/2}$$

where $D(z) = \sqrt{2}/\Gamma(z)$ we use Theorem 3.2. This theorem gives us the beginning of the power series expansion of $\zeta(u, s)$ for s near 0, namely

$$\zeta(s, u) = \frac{1}{2} - u - (\log D(u))s + O(s^2).$$

We know also that $u^{-s} = 1 - s \log u + O(s^2)$ so the term of degree 1 in the power series expansion of $N^{-s} \sum_{j=0}^{N-1} \zeta(s, \{x + j/N\})$ is

$$\sum_{j=0}^{N-1} -\log D\left(\{x + j/N\}\right) - \log N \sum_{j=0}^{N-1} \frac{1}{2} - \left\{ x + \frac{j}{N} \right\}.$$

The addition formula implies that the above has to be equal to the first term in the power series expansion near $s = 0$ of $\zeta(s, \{Nx\})$ which is

$$-\log D\left(\{Nx\}\right).$$

We first evaluate the sum $\sum_{j=0}^{N-1} \frac{1}{2} - \left\{ x + \frac{j}{N} \right\}$. We have

$$\sum_{j=0}^{N-1} \left\{ x + \frac{j}{N} \right\} = \sum_{j=0}^{j'} \{x\} + \frac{j}{N} + \sum_{j=0}^{N-1} \{x\} + \frac{j}{N} - 1$$

$$= N\{x\} + \frac{1}{N} \frac{N(N-1)}{2} - (N - 1 - j')$$

$$= N\{x\} + \frac{(N-1)}{2} - \rho$$

$$= \{Nx\} + \frac{(N-1)}{2}$$

and therefore

$$\sum_{j=0}^{N-1} \frac{1}{2} - \left\{ x + \frac{j}{N} \right\} = \frac{N}{2} - \{Nx\} - \frac{(N-1)}{2} = -\{Nx\} + \frac{1}{2}.$$

We know that $\{Nx\} = Nx - [Nx]$ so we find that

$$-\log N \sum_{j=0}^{N-1} \frac{1}{2} - \left\{ x + \frac{j}{N} \right\} = -\log \frac{N^{[Nx]}}{N^{Nx - \frac{1}{2}}}.$$

Since the logarithm transforms products into sums we see that

$$\sum_{j=0}^{N-1} (-\log D\left(\{x + j/N\}\right)) = -\log \prod_{j=0}^{N-1} D(\{x + j/N\}).$$

Hence the term of degree 1 in the power series expansion near $s = 0$ of the left hand side of the addition formula simplifies to

$$-\log \prod_{j=0}^{N-1} (\{x + j/N\}) - \log \frac{N^{[Nx]}}{N^{Nx - \frac{1}{2}}}.$$

Since this term has to be equal to $-\log D\left(\{Nx\}\right)$ we see that after exponentiation and a few lines of algebra we get the identity

$$\prod_{j=0}^{N-1} (\{x + j/N\}) (N^{[Nx]}) = N^{Nx - \frac{1}{2}} D\left(\{Nx\}\right).$$

We know that $D(z) = \sqrt{2\pi}\,\Gamma(z)$ and the gamma function satisfies $\Gamma(z+1) = \Gamma(z)z$. Using this identity repeatedly we find that

$$D(\{Nx\}) = (Nx-1)(Nx-2)\cdots(Nx-[Nx])D(Nx)$$
$$= N^{[Nx]}(x - \frac{1}{N})(Nx - \frac{2}{N})\cdots(x - \frac{[Nx]}{N})D(Nx)$$

and similarly we find that the product $\prod_{j=0}^{N-1}(\{x + j/N\})$ is equal to

$$\prod_{j=0}^{j'}\left(x + \frac{j}{N} - 1\right)\cdots\left(x + \frac{j}{N} - [x]\right)$$

$$\times \prod_{j=j'+1}^{N-1}\left(x + \frac{j}{N} - 1\right)\cdots\left(x + \frac{j}{N} - [x] - 1\right)\prod_{j=0}^{N-1}D\left(x + \frac{j}{N}\right)$$

$$= \prod_{j=0}^{N-1}\left(x + \frac{j}{N} - 1\right)\cdots\left(x + \frac{j}{N} - [x]\right)\prod_{j=j'+1}^{N-1}\left(x + \frac{j}{N} - [x] - 1\right)$$

$$\times \prod_{j=0}^{N-1}D\left(x + \frac{j}{N}\right)$$

So it suffices to prove

$$\prod_{j=0}^{N-1}\left(x + \frac{j}{N} - 1\right)\cdots\left(x + \frac{j}{N} - [x]\right)\prod_{j=j'+1}^{N-1}\left(x + \frac{j}{N} - [x] - 1\right)$$

$$= \left(x - \frac{1}{N}\right)\left(x - \frac{2}{N}\right)\cdots\left(x - \frac{[Nx]}{N}\right).$$

Collecting terms we see that the above equality holds.

XV.4   Zeta Functions

Exercise XV.4.1. *(a) Show that $\zeta(s)$ has zeros of order 1 at the even negative integers.*
(b) Show that the only other zeros are such that $0 \le \mathrm{Re}\,(s) \le 1$.
(c) Prove that the zeros of (b) actually have $\mathrm{Re}\,(s) = 1/2$. [You can ask the professor teaching the course for a hint on that one.]

Solution. (a) Theorem 4.5 says that

$$\zeta(s) = (2\pi)^s\Gamma(1-s)\frac{\sin(\pi s/2)}{\pi}\zeta(1-s)$$

If s is real and negative, we see from the definitions that $\Gamma(1-s) \ne 0$ and $\zeta(1-s) \ne 0$. Also, $\Gamma(1-s)$ and $\zeta(1-s)$ are holomorphic for $\mathrm{Re}\,(s) < 0$, so since $\sin(\pi s/2)$ has simple zeros at the negative even integers, we conclude that $\zeta(s)$ has simple zeros at the negative even integers.

(b) We first prove that ζ has no zeros for Re $(s) > 1$. For that, we use the fact that

$$\zeta(s) = \prod_p \left(1 - \frac{1}{p^s}\right)^{-1} \quad \text{for Re } (s) > 1$$

where the product is over all prime numbers (see Theorem 1.1 Chapter XVI in Lang's book). Since for Re $(s) > 1$, the sum

$$\sum_p \frac{1}{p^s}$$

converges absolutely, we conclude that the product $\prod_p (1 - \frac{1}{p^s})$, converges, so $\zeta(s) \neq 0$ for Re $(s) > 1$. The fact that the only other zeros of ζ with Re $(s) < 0$ are at the negative even integers follows from what we have just shown and the identity

$$\zeta(s) = (2\pi)^s \Gamma(1 - s)\frac{\sin(\pi s/2)}{\pi}\zeta(1 - s).$$

(c) This is the Riemann hypothesis, one of the big unsolved problem in mathematics.

Exercise XV.4.2. *Define* $F(z) = \xi \left(\frac{1}{2} + iz\right)$. *Prove that* $F(z) = F(-z)$.

Solution. Theorem 4.6 says that $\xi(s) = \xi(1 - s)$, so clearly

$$F(z) = \xi \left(\frac{1}{2} + iz\right) = \xi \left(1 - \frac{1}{2} - iz\right) = \xi \left(\frac{1}{2} - iz\right) = F(-z).$$

Exercise XV.4.3. *Let C be the contour as shown on the figure below.*

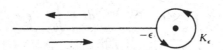

Thus the path consists of $]-\infty, -\epsilon]$, *the circle which we denote by* K_ϵ, *and the path from* $-\epsilon$ *to* $-\infty$. *On the plane from which the negative real axis has been detected, we take the principal value for the log, and for complex s,*

$$z^{-s} = e^{-s \log z}.$$

The integrals will involve z^s, *and the two values for* z^s *in the first and third integral will differ by a constant.*
(a) Prove that the integral

$$\int_C e^z z^{-s} dz$$

defines an entire function of s.

(b) Prove that for Re $(1 - s) > 0$ *we have*

$$\int_C e^z z^{-s} dz = 2i \sin \pi s \int_0^\infty e^{-u} u^{-s} du.$$

(c) Show that

$$\frac{1}{\Gamma(s)} = \frac{1}{2\pi i} \int_C e^z z^{-s} dz.$$

The contour integral gives another analytic continuation for $1/\Gamma(s)$ *to the whole plane.*

Solution. (a) It is clear from the differentiation lemma (Lemma 1.1) and the exponential decay of the integrand, that the integral

$$\int_C e^z z^{-s} dz$$

defines an entire funtion of s.
(b) We can write

$$\int_C e^z z^{-s} dz = \int_{-\infty}^{-\epsilon} + \int_{K_\epsilon} + \int_{-\epsilon}^{-\infty} e^z z^{-s} dz.$$

The integral over K_ϵ goes to 0 as $\epsilon \to \infty$ because for $|z| = \epsilon$,

$$|e^z z^{-s}| \leq C \epsilon^{\text{Re } (s)}$$

hence

$$\left| \int_{K_\epsilon} e^z z^{-s} dz \right| \leq C' \epsilon^{1 - \text{Re } (s)}.$$

So

$$\int_C e^z z^{-s} dz = \int_{-\infty}^0 e^t e^{-s(\log(-t) - i\pi)} dt + \int_0^{-\infty} e^t e^{-s(\log(-t) + i\pi)} dt$$

$$= (e^{i\pi s} - e^{-i\pi s}) \int_0^\infty e^{-u} u^{-s} du$$

$$= 2i \sin \pi s \int_0^\infty e^{-u} u^{-s} du$$

as was to be shown.
(c) Since for Re $(1 - s) > 0$ we have $\Gamma(1 - s) = \int_0^\infty e^{-u} u^{-s} du$ and

$$\Gamma(s)\Gamma(1 - s) = \frac{\pi}{\sin(\pi s)}$$

we see at once from (b) that

$$\frac{1}{\Gamma(s)} = \frac{1}{2\pi i} \int_C e^z z^{-s} dz.$$

XVI
The Prime Number Theorem

XVI.1   Basic Analytic Properties of the Zeta Function

Exercise XVI.1.1. *Let* f *and* g *be two functions defined on the integers* > 0 *and* $\leq n + 1$. *Assume that* $f(n + 1) = 0$. *Let* $G(k) = g(1) + \cdots + g(k)$. *Prove the formula for summation by parts:*

$$\sum_{k=1}^{n} f(k)g(k) = \sum_{k=1}^{n}(f(k) - f(k + 1))G(k).$$

Solution. We define $G(0) = 0$. Then

$$\sum_{k=1}^{n} f(k)g(k) = \sum_{k=1}^{n} f(k)(G(k) - G(k - 1))$$

$$= \sum_{k=1}^{n} f(k)G(k) - \sum_{k=1}^{n} f(k)G(k - 1)$$

$$= \sum_{k=1}^{n} f(k)G(k) - \sum_{m=0}^{n} -1 f(m + 1)G(m)$$

$$= \sum_{k=1}^{n} f(k)G(k) - \sum_{m=1}^{n} f(m + 1)G(m)$$

the last equality holding because $f(1)G(0) = 0 = f(n + 1)G(n)$. Hence

$$\sum_{k=1}^{n} f(k)g(k) = \sum_{k=1}^{n}(f(k) - f(k + 1))G(k)$$

as was to be shown.

Exercise XVI.1.2. *Prove the integral expression for* Φ *in Proposition 1.4.*

Solution. We order the primes in an increasing sequence $2 = p_1 < p_2 < \cdots$. Then we have

$$\int_1^\infty \frac{\varphi(x)}{x^{s+1}}\,dx = \int_1^{p_1} \frac{\varphi(x)}{x^{s+1}}\,dx + \sum_{n=1}^\infty \int_{p_n}^{p_{n+1}} \frac{\varphi(x)}{x^{s+1}}\,dx = \sum_{n=1}^\infty \int_{p_n}^{p_{n+1}} \frac{\varphi(x)}{x^{s+1}}\,dx$$

because $\varphi(x) = 0$ for $x \in (1, p_1)$. We have

$$\int_{p_n}^{p_{n+1}} \frac{\varphi(x)}{x^{s+1}}\,dx = \varphi(p_n) \int_{p_n}^{p_{n+1}} \frac{1}{x^{s+1}}\,dx = \varphi(p_n)\frac{(p_n^{-s} - p_{n+1}^{-s})}{s}.$$

Therefore

$$s\int_1^\infty \frac{\varphi(x)}{x^{s+1}}\,dx = \sum_{n=1}^\infty \varphi(p_n)(p_n^{-s} - p_{n+1}^{-s}).$$

But $\varphi(p_n) = \sum_{j=1}^n \log p_j$ so taking finite sums and summing by parts we find

$$\sum_{n=1}^N \varphi(p_n)(p_n^{-s} - p_{n+1}^{-s}) = \sum_{n=1}^N p_n^{-s} \log p_n.$$

Letting $N \to \infty$ we obtain

$$s\int_1^\infty \frac{\varphi(x)}{x^{s+1}}\,dx = \sum_{n=1}^\infty \frac{\log p_n}{p_n^s} = \Phi(s),$$

as was to be shown.

Exercise XVI.1.3. *Let* $\{a_n\}$ *be a sequence of complex numbers such that* $\sum a_n$ *converges. Let* $\{b_n\}$ *be a sequence of real numbers which is increasing, i.e.,* $b_n \leq b_{n+1}$ *for all n, and* $b_n \to \infty$ *as* $n \to \infty$. *Prove that*

$$\lim_{N\to\infty} \frac{1}{b_N} \sum_{n=1}^N a_n b_n = 0.$$

Does this conclusion still hold if we only assume that the partial sums of $\sum a_n$ *are bounded?*

Solution. Given $\epsilon > 0$, select a positive integer n_0 such that for all $m > n_0$ we have $|\sum_{n=n_0+1}^m a_n| < \epsilon$ and $b_{n_0} \geq 0$. Then for $N > n_0$ splitting the sum we obtain

$$\left| \frac{1}{b_N} \sum_{n=1}^N a_n b_n \right| \leq \left| \frac{1}{b_N} \sum_{n=1}^{n_0} a_n b_n \right| + \left| \frac{1}{b_N} \sum_{n=n_0+1}^N a_n b_n \right|.$$

The first sum will be $\leq \epsilon$ for all large N. For the second sum we use summation by parts to obtain, after some elementary computations,

$$\sum_{n=n_0+1}^N a_n b_n = b_N(A_N - A_{n_0}) - \sum_{k=n_0+1}^{N-1} (A_k - A_{n_0})(b_{k+1} - b_k),$$

where $A_n = \sum_{k=1}^{n} a_k$ are the partial sums. Therefore by the triangle inequality, the fact that $|A_k - A_{n_0}| < \epsilon$ for all $k \geq n_0$ and that $\{b_k\}$ increases we get

$$\left| \sum_{n=n_0+1}^{N} a_n b_n \right| \leq |b_N|\epsilon + \epsilon(b_N - b_{n_0}),$$

hence for all large N we have

$$\left| \frac{1}{b_N} \sum_{n=n_0+1}^{N} a_n b_n \right| \leq 3\epsilon$$

which concludes the proof.

If we only assume that the partial sums of $\sum a_n$ are bounded we cannot conclude that the limit is 0. Indeed, let $a_n = (-1)^n$ and $b_n = 2^n$. Then

$$\frac{1}{b_N} \sum_{n=1}^{N} a_n b_n = \frac{1}{2^N} \frac{-2 - (-2)^{N+1}}{1+2} = \frac{-2}{3 \times 2^N} + \frac{2 \times (-1)^N}{3}$$

and the above expression does not have a limit as $N \to \infty$.

Exercise XVI.1.4. *Let $\{a_n\}$ be a sequence of complex numbers. The series*

$$\sum_{n=1}^{\infty} \frac{a_n}{n^s}$$

*is called a **Dirichlet series**. Let σ_0 be a real number. Prove that if the Dirichlet series converges for some value of s with $\mathrm{Re}(s) = \sigma_0$, then it converges for all s with $\mathrm{Re}(s) > \sigma_0$, uniformly on every compact subset of this region.*

Solution. Let s_0 denote the point where the Dirichlet series converges. Let K be a compact subset of the half plane $\mathrm{Re}(s) > \sigma_0$. Suppose $s \in K$. To simplify the notation we define $\alpha_n = \log n$. To show that the series $\sum a_n/n^s$ converges uniformly on K it is sufficient to show that it is uniformly Cauchy on K. We write

$$\sum_{k=m}^{n} \frac{a_k}{k^s} = \sum_{k=m}^{n} a_k e^{-s\alpha_k} = \sum_{k=m}^{n} a_k e^{-s_0\alpha_k} e^{-(s-s_0)\alpha_k} = \sum_{k=m}^{n} \frac{a_k}{k^{s_0}} e^{-(s-s_0)\alpha_k}.$$

Let $c_k = a_k/k^{s_0}$, $b_k = e^{-(s-s_0)\alpha_k}$ and $S_k = \sum_{j=1}^{k} c_j$. Summing by parts we find

$$\sum_{k=m}^{n} \frac{a_k}{k^{s_0}} e^{-(s-s_0)\alpha_k} = \sum_{k=m}^{n} c_k b_k = \sum_{k=m}^{n-1} S_k(b_k - b_{k+1}) - S_{m-1}b_m + S_n b_n.$$

Putting absolute values and using the triangle inequality, we see that we must estimate the three terms

$$\left| \sum_{k=m}^{n-1} S_k(b_k - b_{k+1}) \right|, \quad |S_{m-1}b_m| \quad \text{and} \quad |S_n b_n|.$$

Since S_k converges as $k \to \infty$, there exists a positive number M such that $|S_k| \leq M$ for all k. Hence

$$\left| \sum_{k=m}^{n-1} S_k(b_k - b_{k+1}) \right| \leq M \sum_{k=m}^{n-1} |b_k - b_{k+1}|.$$

Let $z = s - s_0$, so that

$$b_k - b_{k+1} = e^{-z\alpha_k} - e^{-z\alpha_{k+1}} = z \int_{\alpha_k}^{\alpha_{k+1}} e^{-zt} \, dt.$$

But there exists $\delta, B > 0$ such that if $z = s - s_0 = x + iy$, then $x \geq \delta > 0$ and $|z| \leq B$ uniformly for $s \in K$. So

$$|b_k - b_{k+1}| \leq |z| \left| \int_{\alpha_k}^{\alpha_{k+1}} e^{-zt} \, dt \right| \leq B \int_{\alpha_k}^{\alpha_{k+1}} e^{-xt} \, dt \leq \frac{B}{\delta}(e^{-x\alpha_k} - e^{-x\alpha_{k+1}})$$

and therefore

$$\left| \sum_{k=m}^{n-1} S_k(b_k - b_{k+1}) \right| \leq C(e^{-x\alpha_m} - e^{-x\alpha_n})$$

$$\leq Ce^{-\delta\alpha_m}$$

so the term $\sum_{k=m}^{n-1} S_k(b_k - b_{k+1})$ is uniformly Cauchy as $n, m \to \infty$. The inequalities

$$|S_{m-1}b_m| \leq M|e^{-z\alpha_m}| \leq Me^{-\delta\alpha_m}$$

and

$$|S_n b_n| \leq Me^{-\delta\alpha_n}$$

combined with the estimate for $\sum_{k=m}^{n-1} S_k(b_k - b_{k+1})$ show that

$$\left| \sum_{k=m}^{n} \frac{a_k}{k^s} \right| \to 0 \quad \text{as } n, m \to \infty$$

uniformly for $s \in K$.

Exercise XVI.1.5. *Let $\{a_n\}$ be a sequence of complex numbers. Assume that there exists a number C and $\sigma_1 > 0$ such that*

$$|a_1 + \cdots + a_n| \leq Cn^{\sigma_1} \quad \text{for all } n.$$

Prove that $\sum a_n/n^s$ converges for $\mathrm{Re}(s) > \sigma_1$. [Use summation by parts.]

Solution. Let $\sigma = \mathrm{Re}(s)$ and $S_k = a_1 + \cdots + a_k$. Summing by parts we find

$$\sum_{k=m}^{n} \frac{a_k}{k^s} = \sum_{k=m}^{n-1} S_k \left(\frac{1}{k^s} - \frac{1}{(k+1)^s} \right) - \frac{S_{m-1}}{m^s} + \frac{+S_n}{n^s}.$$

Using the same estimate as in Theorem 1.2 (the mean value theorem) we get

$$\left| \frac{1}{k^s} - \frac{1}{(k+1)^s} \right| \leq \frac{|s|}{k^{\sigma+1}},$$

so if $\sigma - \sigma_1 = \epsilon$, then

$$\left| \sum_{k=m}^{n-1} S_k \left(\frac{1}{k^s} - \frac{1}{(k+1)^s} \right) \right| \leq C|s| \sum_{k=m}^{n-1} \frac{1}{k^{1+\epsilon}}.$$

But $\sum 1/k^{1+\epsilon}$ converges and we have

$$\left| \frac{S_{m-1}}{m^s} \right| \leq \frac{C}{m^{\epsilon}} \quad \text{and} \quad \left| \frac{S_n}{n^s} \right| \leq \frac{C}{n^{\epsilon}}$$

so we conclude that $\sum a_k/k^s$ satisfies the Cauchy criterion of convergence.

Exercise XVI.1.6. *Prove the following theorem.*

Let $\{a_n\}$ be a sequence of complex numbers, and let A_n denote the partial sum

$$A_n = a_1 + \cdots + a_n.$$

Let $0 \leq \sigma_1 < 1$, and assume that there is a complex number ρ such that for all n we have

$$|A_n - n\rho| \leq Cn^{\sigma_1},$$

or in other words, $A_n = n\rho + O(n^{\sigma_1})$. Then the function f defined by the Dirichlet series

$$f(s) = \sum \frac{a_n}{n^s} \quad \text{for } \operatorname{Re}(s) > 1$$

has an analytic continuation to the region $\operatorname{Re}(s) > \sigma_1$, where it is analytic except for a simple pole with residue ρ at $s = 1$.

[Hint: Consider $f(s) - \rho\zeta(s)$, use Theorem 1.2, and apply Exercise 5.]

Solution. The hint gives the proof away. Applying Exercise 5 to the Dirichlet series

$$\sum \frac{a_n - \rho}{n^s} = \sum \frac{a_n}{n^s} - \rho \sum \frac{1}{n^s}$$

we see that this series converges uniformly on compact subsets of the half plane $\operatorname{Re}(s) > \sigma_1$ so it defines a holomorphic function there. Since the zeta function is holomorphic on $\operatorname{Re}(s) > \sigma_1$ except for a simple pole at $s = 1$ we are done.

XVI.2   The Main Lemma and its Application

Exercise XVI.2.1. *Prove the lemma allowing you to differentiate under the integral sign in as great a generality as you can, but including at least the case used in the case of the Laplace transform used before Lemma 2.2.*

Solution. Suppose f is bounded and piecewise continuous. We use the notation of the differentiation lemma, §1 of Chapter XV. Let $f(t, z) = f(t)e^{-zt}$. Let U denote the right half plane $\operatorname{Re}(z) > 0$ and let $I = [0, \infty)$. Finally, let K be a compact subset of U. There exists $\delta > 0$ such that for all $z \in K$ we have $\mathrm{Re}(z) > \delta$. Observe that if B is a bound for f, then for all $z \in K$ we have

$$|f(t, z)| \le Be^{-\delta t}$$

so $\int_I f(t, z)dt$ is uniformly convergent for $z \in K$, and for each t the function $z \mapsto f(t, z)$ is analytic. Let $\{I_n\}$ be a sequence of closed intervals increasing to I. Then choosing a disc D as in the differentiation lemma, we find that

$$f(t, z) = \frac{1}{2\pi i} \int_\gamma \frac{f(t, \zeta)}{\zeta - z} d\zeta$$

so

$$g(z) = \frac{1}{2\pi i} \int_I \int_\gamma \frac{f(t, \zeta)}{\zeta - z} d\zeta \, dt.$$

For each n, define

$$g_n(z) = \frac{1}{2\pi i} \int_{I_n} \int_\gamma \frac{f(t, \zeta)}{\zeta - z} d\zeta \, dt,$$

so that restricting z as in the differentiation lemma we get

$$g_n(z) = \frac{1}{2\pi i} \int_\gamma \frac{1}{\zeta - z} \left[\int_{I_n} f(t, \zeta)dt \right] d\zeta.$$

The expression in bracket is continuous in ζ, so g_n is analytic. The hypotheses imply that $g_n \to g$ uniformly on K, so g is analytic, as was to be shown.

Remark. If we only assume that f is bounded and that $\int_a^b |f(t)|dt$ exists for all $a, b \ge 0$ the only difficulty is to show that the expression $\int_{I_n} f(t, \zeta)dt$ is continuous in ζ. Writing $I_n = [a, b]$ we have

$$\left| \int_{I_n} f(t, \zeta)dt - \int_{I_n} f(t, \zeta_0)dt \right| \le \int_a^b |f(t)| \, |e^{-\zeta t} - e^{-\zeta_0 t}| dt.$$

Uniform continuity of a continuous function on a compact set implies that given $\epsilon > 0$ there exists $\delta > 0$ such that if $|\zeta - \zeta_0| < \delta$, then

$$|e^{-\zeta t} - e^{-\zeta_0 t}| < \epsilon \quad \text{for all } t \in [a, b].$$

So

$$\left| \int_{I_n} f(t, \zeta)dt - \int_{I_n} f(t, \zeta_0)dt \right| < \epsilon \int_a^b |f(t)|dt$$

whenever $|\zeta - \zeta_0| < \delta$.